Australia's

Biodiversity

Living Wealth

Andrew J. Beattie
Editor

Foreword Paul R. Ehrlich

Authors
Andrew Beattie
David Briscoe
Frank Burrows
Lucy Crowley
Peter Fairweather
Dinah Hales
Lesley Hughes
Jim Kohen
George McKay
Malcom Reed
Louise Rodgerson
Patricia Selkirk
Noel Tait
Duncan Veal
Mark Westoby

Executive Editor
Andrew Beattie

Managing Editor
Fiona Doig

Editorial Assistant
Robyn Delves

Illustrators
Nicola Oram (cover, pages 15 and 40)
Christine Turnbull

Design and Production
Watch This! Design, Sydney

A few words about the Research Unit: Macquarie University's Research Unit for Biodiversity and Bioresources was created to explore the biological wealth of Australia and to develop new ways of saving, understanding and sustainably using it. Members include botanists and zoologists, ecologists, microbiologists, chemists and biotechnologists working on multidisciplinary approaches to the conservation and management of the nations's biological heritage.

COVER *Size and scale in biodiversity. In this scene the wren and the violet are giants because most species in their environment are smaller. Some of these species can be seen here but the majority (including many invertebrates, most fungi and all bacteria) are minute and are invisible without a microscope.*

Biodiversity, Australia's Living Wealth has been published with the assistance of the Commonwealth Department of the Environment, Sport and Territories. The views expressed in this book are not necessarily those of the Commonwealth.

For Information on Commonwealth biodiversity programs please call the department's Community Information Unit on (008) 803 772

Other biodiversity publications available from the department include:
Biodiversity Series, Paper No. 1
Biodiversity and its value.
Biodiversity Series, Paper No. 2
Australia's biodiversity: an overview of selected significant components.

First published 1995 by
Reed Books
A part of Reed Books Australia
Level 9, North Tower
1-5 Railway St., Chatswood, NSW 2067

in association with
Macquarie University
Sydney NSW 2109

National Library of Australia
Catalogue-in-Publication data

Biodiversity: Australia's living wealth
Bibliography
Includes index
ISSBN 0 7301 0482 6

1. Biological diversity
2. Biological diversity-Australia
I. Beattie, Andrew J (Andrew James), 1943-.

575

Printed in Australia on
total chlorine free paper by
Griffin Press, Netley, South Australia.

The Purpose of This Book

The United Nations Conference on Environment and Development (the "Earth Summit") was held in Rio de Janeiro during 1992. Australia signed the Convention on Biological Diversity at the Conference and ratified it on 18 June 1993. The Convention was developed in recognition of the present and future value of biological diversity and its significant reduction around the world. Through the Convention, Australia and many other countries are involved in an international partnership to halt the global loss of biodiversity. The Convention came into force on 29 December 1993.

It has all seemed very sudden—indeed the science of biodiversity is very new and most of us are unfamiliar with its vocabulary. This has raised the spectre that it will become merely another rallying point for the extremists in the environmental debate: a saviour for those who would deny development of natural resources and doom to those for whom biodiversity conservation suggests the end of their livelihood.

It is essential that we find ways to balance the development and conservation of our natural resources. The purpose of this book is to introduce the science of biodiversity in the hope that increased understanding of its foundations will lead to informed debate and a wider consensus.

This book is written particularly for those without a formal education in biology but who are either interested in or involved with making government and business decisions that influence the environment. It will inform you on how biodiversity is measured, how it evolves, what affects it and why it is important. There are also special sections on Australia's unique biodiversity and the exciting scientific challenges it presents.

This book revolves around four fundamental principles:

• *Biodiversity is crucial to environmental management.*
Natural ecosystems and their occupants provide us with effective waste disposal, clean air and water, productive soils, and the control of pests, disease, floods and erosion.

• *Biodiversity is biological wealth.*
Some of our most important industries are rooted in biodiversity—including agriculture, forestry, fisheries and tourism. Biodiversity provides resources as diverse as medicines, construction materials, textiles, foods of every conceivable kind and a vast range of potentially valuable products, the scope of which we are only just beginning to imagine.

• *Conservation of biodiversity is a challenge for us all.*
All users and consumers of our natural resources have a vested interest in the conservation and management of biodiversity.

• *Biodiversity is a frontier.*
The majority of organisms are tiny and the variety of living things on Earth is so vast that to find them all presents a challenge to science and technology as great as exploring other planets. The need to assess and understand biodiversity will make this an age of exploration and discovery.

Foreword
While There's Time

I first came to Australia in 1965 to spend a sabbatical leave with Professor Charles Birch, one of the world's most distinguished ecologists. That was when I became acquainted with the continent's unique stock of biodiversity and Australia's superb environmental scientists, then as now probably the finest per capita in the world.

Australia's isolation not only gave it a rich and unique flora and fauna, but also somewhat retarded its destruction. For about 50,000 years, the original Australians trod lightly on their continent. The stunning expansion of human numbers that, along with skyrocketing consumption, now threatens the planet's life support systems has not been as severe in Australia as elsewhere. Nonetheless, a relatively small number of representatives of industrial civilisation have managed to destroy the integrity of many of the continent's ecosystems because those systems were especially vulnerable to disruption. Australia's biological riches are in grave danger, and along with them the prosperity of future generation. Australians must recognise and reject those policies and life-styles that are destructive to the environment and work together to protect their biodiversity.

Fortunately, there is a broad-based movement among Australians to preserve their natural heritage and secure the future of their nation. I was greatly encouraged two years ago to meet a wonderful group of farmers in the Tammin area of Western Australia who were busily replanting native vegetation around wheatfields. They were both restoring salinized areas to production and creating habitat that would help perpetuate the severely threatened biodiversity of the region. But, of course, I was saddened to realise that for every tree or shrub the farmers of Tammin managed to plant, hundreds or thousands were simultaneously being destroyed elsewhere in Australia.

Today, it is a matter of urgency that everyone in Australia including those in government, business, biodiversity-based industries and education understands the meaning and importance of biodiversity. If this is achieved, then the Sydney Olympics in the year 2000 may be held in a country where the balance between conservation and development serves as a model for the rest of the world. This splendid book will be a very important aid in these crucial tasks.

—Professor Paul R. Ehrlich,
Bing Professor of Population Studies,
Stanford University, California.

Preface

Australia has a living heritage that rivals any in the world. Many species of our various and unusual plants and animals, and the natural communities in which they are found, occur only on this continent. The figures are astonishing: 90 per cent of our 20,000 named species of flowering plants are unique to this country. Add to this half the marsupial species of the world, about half of our 770 bird species and over 80 per cent of our 700 reptile species. Australia also has at least 4,000 species of fish. There are innumerable insects, molluscs, fungi and other organisms that science has not yet fully explored.

Australia is one of only 12 megadiverse nations, which, altogether, harbour between 60 and 70 per cent of all known species. This is recognised not only by scientific organisations but also by international institutions such as the World Bank and World Resources Institute. Australian megadiversity contains both a remarkably large number of species and a very high proportion of species found nowhere else on Earth.

Unfortunately, like most cultures, we are faced with an increasing loss of biodiversity, both in terms of species and natural communities. The most obvious extinctions in Australia include 20 mammal, 10 bird and 97 plant species, all in the past two centuries.

Extinctions among other groups of organisms are less well understood. This is especially crucial as the mammals, birds and flowering plants make up less than 20 per cent of the species in any natural community. The bulk of biodiversity is found among the other, less-familiar animals and plants and the micro-organisms: insects, mites, molluscs, worms, single-celled animals, algae, mosses, fungi and a vast array of bacteria. Within these groups, many species are likely to disappear before they are discovered.

Edward O. Wilson, one of the great biologists of the world, calls the loss of species worldwide 'the silent haemorrhage'. This stark metaphor highlights the increasing pace and magnitude with which the life of the planet is ebbing away.

The limits to our knowledge of biodiversity were emphasised by a catalogue of the organisms of the Earth, *The five kingdoms*, by Lynn Margulis and Karlene Schwartz (1988). It was commissioned by the National Aeronautics and Space Administration (NASA) in the USA to help prepare missions looking for life on Mars. As new and more bizarre life forms were discovered on Earth it became clear that to all but the specialist many organisms looked alien. Ironically, cataloguing life on Earth was done to ensure it could be distinguished from any life that might be found on Mars. More ironically we continue to spend billions of dollars looking for life in the cosmos when we understand so little about the life on our own planet.

Against this background of extinctions and limited knowledge, we still experience the thrill of discovery here on Earth. Thousands of new species are described each year, including many unexpected finds: since 1988 four new species of monkeys have been discovered. They include the Sun-tailed Guenon in Gabon and the Maues Marmoset in Brazil. Off the coast of Peru, unlikely as it may seem, a new species of whale has been described. An Indonesian island has revealed a giant bee, four centimetres long, with huge sabre-like jaws, that nests only with termites. The Great Barrier Reef in Australia has produced many new species in recent years but one of the most bizarre is a giant bacterium, visible to the naked eye, which lives in the intestine of surgeon fish.

As we near the end of the twentieth century, many nations are overpopulated and overcrowded. Their landscapes are almost entirely modified by human activity and they suffer widespread pollution and environmental degradation. In contrast, Australia is one of relatively few nations that still have the opportunity to get things right. Although much of our environment is damaged, the balance between human population and ecosystem degradation is not yet as critical as in many other places in the world. Australia can still get the people—environment—economy equation balanced for a productive and sustainable future.

Put another way, we still have the opportunity to get the three phases of biodiversity exploration right:

- *To save it*
- *To understand it*
- *To sustainably use it.*

This task is one for all Australians, for two reasons. First, because there is so much biodiversity there are still numerous species to be discovered. Even large animals like mammals, fish and frogs are being discovered in this country. Discovering a new mammal has all the excitement of discovering a new planet. When it comes to the smaller organisms, almost any place—from backyards to forests, beaches and rock pools—may harbour species new to science. The exploration of biodiversity is still in its infancy and anyone, with a little knowledge and a lot of hard work, can participate.

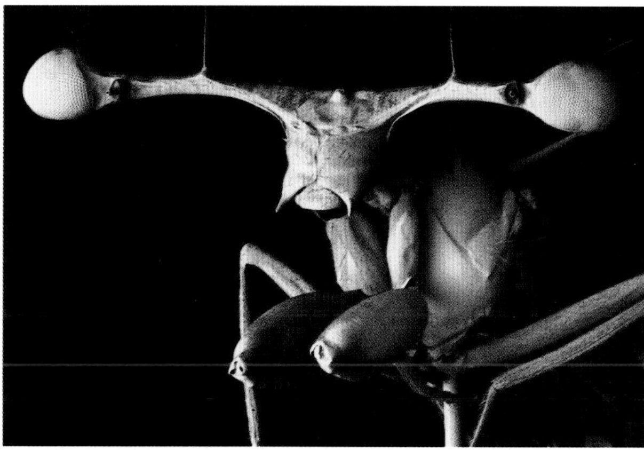

Life on Earth can appear alien: a stalk-eyed fly from Queensland.

Second, Australia and its surrounding seas cannot be devoted exclusively to the conservation of biodiversity. Agriculture, forestry, fishing, tourism, and industry, must remain productive for the economy to be strong. Currently, most of the land and coastal water is available for production, placing Australian biodiversity directly in the hands of everyone that has access to it—whether in a garden, farm, forest, factory, mine, tourist resort, sports ground or crown land. This means that most Australians will have an impact on the exploration, conservation and management of our biodiversity in the future.

—Andrew Beattie

Executive Editor

Research Unit for Biodiversity and Bioresources

Contents

What is Biodiversity?

Biodiversity is a new term, one not included in any dictionary until very recently. It is actually a contraction of the words 'biological diversity'. In its most general sense, it means the grand total of all the profusion of life forms on Earth. It is something we are all very familiar with, only the word is new. At a personal level, we encounter biodiversity every time we go into a garden, eat a meal, take a holiday or go shopping.

Encountering Biodiversity

The masses of produce in a greengrocer store—apples, oranges, bananas, carrots, capsicum, cauliflower, beetroot and onions—are collectively an example of biodiversity. Each of these fruits and vegetables belongs to a distinct biological form called a species. However, what we see is only a tiny fraction of the true biodiversity of the store. On and inside the fruits and vegetables, in the dust on the floor and in the air, exists a myriad of

LEFT *The Great Barrier Reef is about 2,000 kilometres in length, constructed by tiny coral-building animals. Coral reefs are teeming with life and exhibit high biodiversity. However, biodiversity is also extremely important in areas beyond the tropics. These cushion plants (above) live in the salty windswept coastal environment of sub-Antarctic islands.*

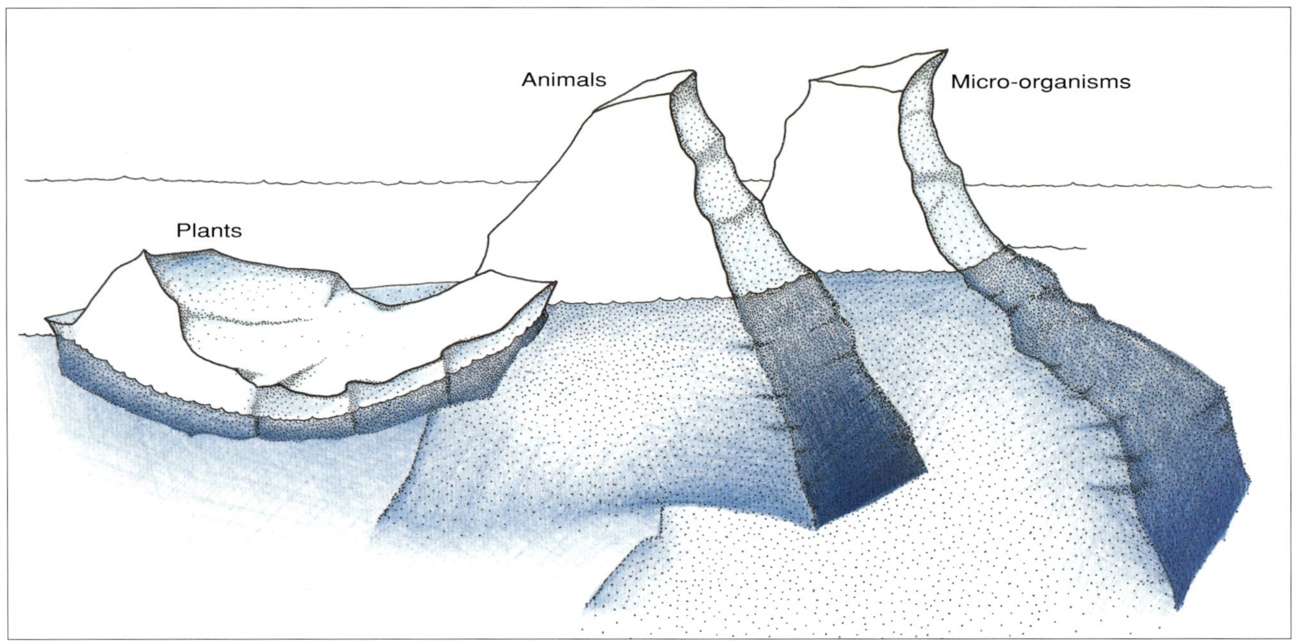

These icebergs represent three major groups—plants, animals and micro-organisms. Above the water shows the proportions of species that are known, while below the waterline lie the proportions that are unknown. The plants are quite well-known compared to the animals and micro-organisms.

species of microscopic animals, bacteria and fungi. Despite being hard to see, they outnumber the vegetable species by thousands to one.

Biodiversity is rather like an iceberg: only a small part is immediately visible but as we explore deeper, far more is revealed. When we think of 'life', familiar things like plants and animals come to mind. It seems obvious that these belong to two separate groups, classified by biologists as kingdoms. However, like the iceberg, the part of biodiversity we cannot see is actually the largest section. Mostly hidden from the naked eye are three other major groups of organisms, each of which is so different to the plants and animals, as well as to each other, that they have their own kingdoms. These are the fungi, the bacteria, and a sometimes bizarre array of organisms some of which are like plants while others are closer to animals. This array includes a great many species that do not fit readily into any other kingdom, largely because they are not well known. To make the checklist of living things look tidy, many scientists lump everything that does not fit neatly into the other four kingdoms under this fifth kingdom. Consequently it is not as organised as the other four and is like a 'too-hard basket' of biological classification.

In order to deal with the enormous diversity of life on Earth, it is not enough

What appears to be one species may actually be several. Close inspection of this longicorn beetle reveals parasitic mites on its surface.

The Fifth Kingdom

The exquisite singled-celled Foraminifera.

Biologists continue to argue over what to do with the organisms that do not fit into the four better-known kingdoms (animals, plants, fungi and bacteria) but to date there has been no satisfactory solution. One scheme that has received much attention in recent years is the creation of a fifth kingdom with the tongue-twisting scientific name Protoctista. This is a very large kingdom of mostly microscopic organisms with features that make them hard to fit elsewhere. It includes organisms such as slime moulds, which are like animals because they move and eat other organisms, produce spores rather like fungi and contain cellulose in their cell walls like plants. These are not rare organisms—most backyards have them.

Many protoctists have no familiar names but are very common. One species of dinoflagellate is a single carnivorous cell with a whip-like tentacle that captures its food. Sailors know it well as it is bioluminescent and, when present in large numbers, makes ocean waves glow in the dark. Euglenoids, single-celled organisms that are common in ponds, also have a tentacle or, more correctly, a flagellum, and move around rapidly in the water. However, they are green and photosynthesise like plants. To complicate things further, under certain conditions, they lose their green colour, stop photosynthesising, and become carnivores. Some protoctists cause animal diseases, such as sleeping sickness, and plant diseases like potato blight. The exquisite miniature shells of the single-celled Foraminifera and their relatives make up a major part of some ocean sediments. In various parts of the world these sediments have become rock and, in ancient times, the Egyptians mined them to build the pyramids.

One of the chief controversies about the fifth kingdom is that it includes green algae, which are also included in the plant kingdom because they manufacture their own food using the energy from sunlight in the process called photosynthesis. These form the familiar slime in ponds, green blushes on tree trunks and the slippery surface on wet paths.

The debate on how many kingdoms there are and how organisms should be divided up among them continues to rage. It is a bit like all the junk that accumulates in a garage. It needs cleaning out and organising, but this is a daunting task because we do not know where to put it or even how to sort it out. Further exploration of biodiversity may reveal a continuum of living things rather than a series of pigeonholes into which everything can be neatly fitted.

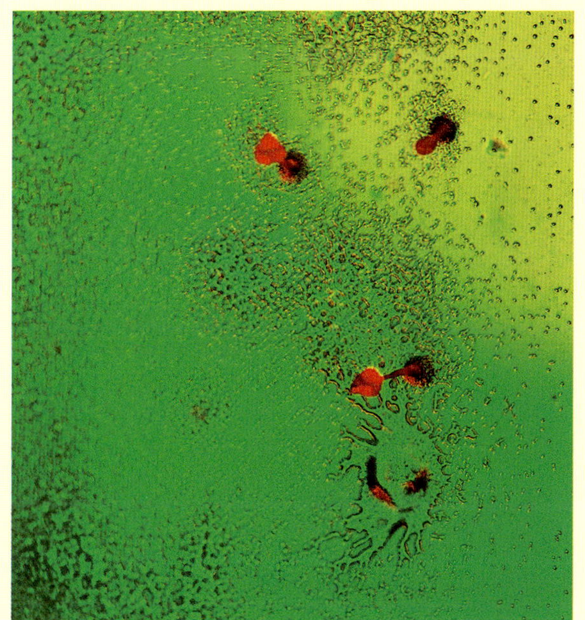

Some organisms like slime moulds are difficult to classify because they have characteristics of animals, plants and fungi, all of which belong to separate kingdoms. Yet slime moulds are common, occurring in leaf litter and rotting logs. Coloured illumination has been used to show up the texture.

to understand it just on the level of kingdoms. Scientists have to examine diversity at many other levels between kingdom and species. For example, no botanist can study all of the organisms within the plant kingdom. They would have to specialise

Some plant biodiversity is dependent upon the Cassowary which is an important seed disperser in tropical forests.

on groups within this kingdom, maybe on flowering plants, then on trees within the group of flowering plants, perhaps eucalypts within the trees and maybe just a handful of particular species.

But this basic framework only shows how biodiversity is classified into groups. It does not help us understand how to count biodiversity if we asked 'how many types—or species—of living things are there?' It does not clarify the complex characteristics that make up each species or each individual. Nor does it explain the diverse interactions between groups of coexisting species. To include these important questions in the equation, we must break the concept of biodiversity down into three more manageable levels: species diversity, genetic diversity and ecosystem diversity.

Species Diversity

As a general rule, a species is a group of individuals that can interbreed with one another, to produce viable offspring, but cannot interbreed with individuals from any other groups. In some cases, very different looking living things, such as domestic dogs, can interbreed. Therefore they belong to the same species. In other cases, superficially very similar-looking organisms may belong to quite different species. For example, the cloud of small brown flies often seen hovering above the fruit in the greengrocer store may be composed of several species.

The species is the most practical unit for measuring biodiversity. This is because each species has its own unique combination of genetic variations that makes it different to all other species. Scientists now have an array of methods for detecting whether a collection of specimens is composed of one, several or many species. We can therefore measure biodiversity by identifying and counting the number of species in a given area. So at this level, biodiversity is the variety of species found.

Counting the number of species on Earth would seem to be a large but straightforward exercise. It is, in fact, a gargantuan task. Imagine an area of tropical rainforest. We can see the trees and

RIGHT *A visitor to the scene at the top would be able to find 100–200 species without too much difficulty. These might include a butterfly and a mangrove on land, and a crab and sea slug in the sea. If everything were magnified 50 times, a large array of smaller species would become apparent. These are shown in the middle of the pyramid and include a nematode worm, paddle worm, comb jelly, arrow worm, jellyfish, tunicate and seaweed. At this magnification, a total of 1,000–5,000 species could be counted. However, if everything was magnified 100–500 times, a vast array of tiny species would be discovered including filamentous algae, single-celled animals and plants, fungi and bacteria such as those shown at the base of the pyramid. No complete counts of these kinds of organisms have ever been done for a single location but 10,000 species is a reasonable estimate.*

The Hidden Dimensions of Biodiversity

many of the smaller plants, and perhaps a few birds, reptiles and some of the more spectacular insects. These easily visible species are the tiny minority. If we look more closely we find that each tree is home to many other species of plants: algae, mosses, ferns, vines and maybe bromeliads or orchids. Many insects live in and on these plants, eating the wood, flowers, fruits, stems and roots. Some, like termites, in turn harbour a variety of tiny animals called protozoans living inside their intestines. The plants, birds, frogs and lizards—indeed all the species in the forest—will each harbour at least one parasitic species; many will be afflicted with several. So when we look at an individual tree, we are not seeing one organism but an entire community made up of ten, 100 or even 1,000 species. The forest floor harbours thousands more species of animals, fungi and bacteria, mostly microscopic. The challenge of discovering how many species there are at a particular place or even in the entire world is detailed in Chapter 8.

The Vast Genetic Library

Biodiversity is like a library. The species are the books and they contain vast amounts of genetic information. Libraries are useful because their books contain information that may be of immediate use, important in the future or simply contribute to the cultural wellbeing of society. But in the biodiversity library, vast quantities of books are hidden, others are in storage waiting to be catalogued, while only a small proportion have been read.

How are Species Described and Named?

Scientists who describe and give names to newly discovered species are taxonomists. Much of their information is gained by studying collections of specimens, mostly stored in herbaria and museums. Plant specimens are pressed, dried and mounted on paper sheets; small animal specimens are often preserved in alcohol, while skeletons and skins of vertebrates are kept in special cabinets. Details of where and when each specimen was collected are recorded, often on computerised databases. Because collections must represent as many species as possible, they are often huge, such as the herbarium at the Royal Botanic Gardens in Sydney, which holds about one million specimens, and the Australian National Insect Collection in Canberra, which holds around 10 million.

A taxonomist working on a new group of organisms will first assemble as many of the known specimens in that group as possible—from institutions all around the world. Specimens are reviewed and then sorted into groups corresponding to species. This involves making judgments about just how much variation is natural between individuals within a species, and how much represents differences between species.

The taxonomist then gives formal names to newly recognised species, along with a description that emphasises attributes most useful to distinguish between species. To be accepted throughout the community of scientists, the names and descriptions must be published in one of a limited number of scientific journals, which subject manuscripts to peer review by other experts before accepting them for publication.

Take as an example the Eastern Grey Kangaroo *(Macropus giganteus)*. While kangaroos are very familiar to Australians, it is necessary to inform scientists from all over the world precisely how this species is related, not only to other kangaroos, but to all other living things as well. It is classified as follows—species: *giganteus;* genus:

Internal parasites can have major effects on their hosts. Populations of the Gouldian Finch, once plentiful in northern Australia, have declined drastically since the turn of the century. A prime suspect is a mite that affects the breathing efficiency of heavily infected birds.

Macropus; family: Macropodidae; order: Marsupialia; class: Mammalia; phylum: Chordata; and kingdom: Animalia. This list is like scientific shorthand. If we read it backwards, it tells us this kangaroo is an animal (Animalia) with a type of spine (Chordata), it has fur and suckles its young (Mammalia) in a pouch (Marsupialia) and belongs to the family of kangaroos sharing certain features (Macropodidae), a few of which appear so closely related that they have immediate common ancestry *(Macropus)* and, finally, specifies one particular kind that breeds only with others sharing the same highly specific set of traits *(giganteus).*

Genetic Diversity

Biodiversity exists within species as well as between species, at the level of genetic diversity. Within almost all species there exists extensive inherited variation that makes each individual organism genetically unique. This variation is important in several ways. In the short term, subtle differences among individuals or populations within a particular species may allow them to adapt to, and therefore exploit, a variety of different environments. For example, yellow and brown striped snails are well camouflaged in long grass, while brown snails are difficult for predators to detect in a woodland environment. Not surprisingly, many species of snails have genetic variations that make some individuals striped and others brown, allowing each species to exploit both environments.

In the longer term, the ability of a species to adapt to environmental change depends on the genetic variability it possesses. Without genetic variability a species cannot evolve and could possibly become extinct when conditions changed. A type of 'forced' evolution has been imposed by humans on domesticated animals and plants. For

example, all domestic dog breeds belong to a single species, *Canis familiaris*. The genes within our first fireside companions were so varied that we have been able to select for breeds as different as the Corgi, Blue Heeler and Great Dane. Likewise, improvements in the productivity of our crops and livestock have been the result of our ability to breed selectively from individuals with desirable genetic variations.

Ecosystem Diversity

Groups of species interact with each other and with their physical environment to form ecosystems. Because the surface of the Earth comprises a huge variety of different environments, diversity exists in the type and variety of ecosystems: ponds, rivers, lakes and swamps; deserts, forests, heaths and grasslands; seashores, estuaries, coral reefs and deep oceans are just a few. Communities of species do not just live in these places—they are an integral part of their environment and create the ecological processes that maintain the system. These processes are the ways in which species use energy, water, gases (such as oxygen and carbon dioxide), mineral elements (such as sulphur and calcium), and inorganic nutrients (like phosphates and nitrates).

Within an ecosystem, interactions between species may appear to be simple and straightforward. For example, the predator–prey relationship seems clear when birds eat ants. However, both birds and ants are involved with many other species. Lizards, for example, also eat ants and are, therefore, in a competitive relationship with birds when ants are scarce. Ants can be predators, too, eating caterpillars and other insect larvae. However, they also disperse the seeds of many different kinds of plants to places safe from predators or fire. So even this apparently simple example involves birds, ants, lizards, caterpillars and many plant species.

Additionally, the birds, ants and lizards are involved in relationships with their parasites, which find shelter and food on or in the bodies of their hosts. These include animals such as ticks and

LEFT *A colourised scanning electron micrograph of house dust reveals a hidden world: filaments of a spider web are shown in grey, a bristle from the seed of a dandelion has been tinted green; the remains of a springtail are picked out in blue while its legs are coloured rusty brown. The fawn colour highlights scales of human skin.*

Biodiversity at a Glance

Biodiversity is the variety of all living things, including plants, animals and micro-organisms.

The Genetic Component: ***Millions of different genes combine in a variety of ways to make distinctive kinds of organisms. The number of possible combinations is so large that every individual is unique. Groups of individuals sharing very similar combinations are often recognisable as breeds, races, varieties, ecological variants or local populations within a single species.***

The Species Component: ***When a combination of genes is shared by a group of individuals that breed predominantly with each other, we call that group a species. One of the great mysteries of modern biology is how many different true-breeding groups exist; that is, how many species are there in the world? The latest count is 1.82 million, but educated guesses range from 10–100 million.***

The Ecosystem Component: ***Species occupying a particular area of land or water interact with each other and their physical environment to form major natural units called ecosystems. The different kinds of environments, combined with the many different species adapted to live in each, generates an enormous diversity of ecosystems.***

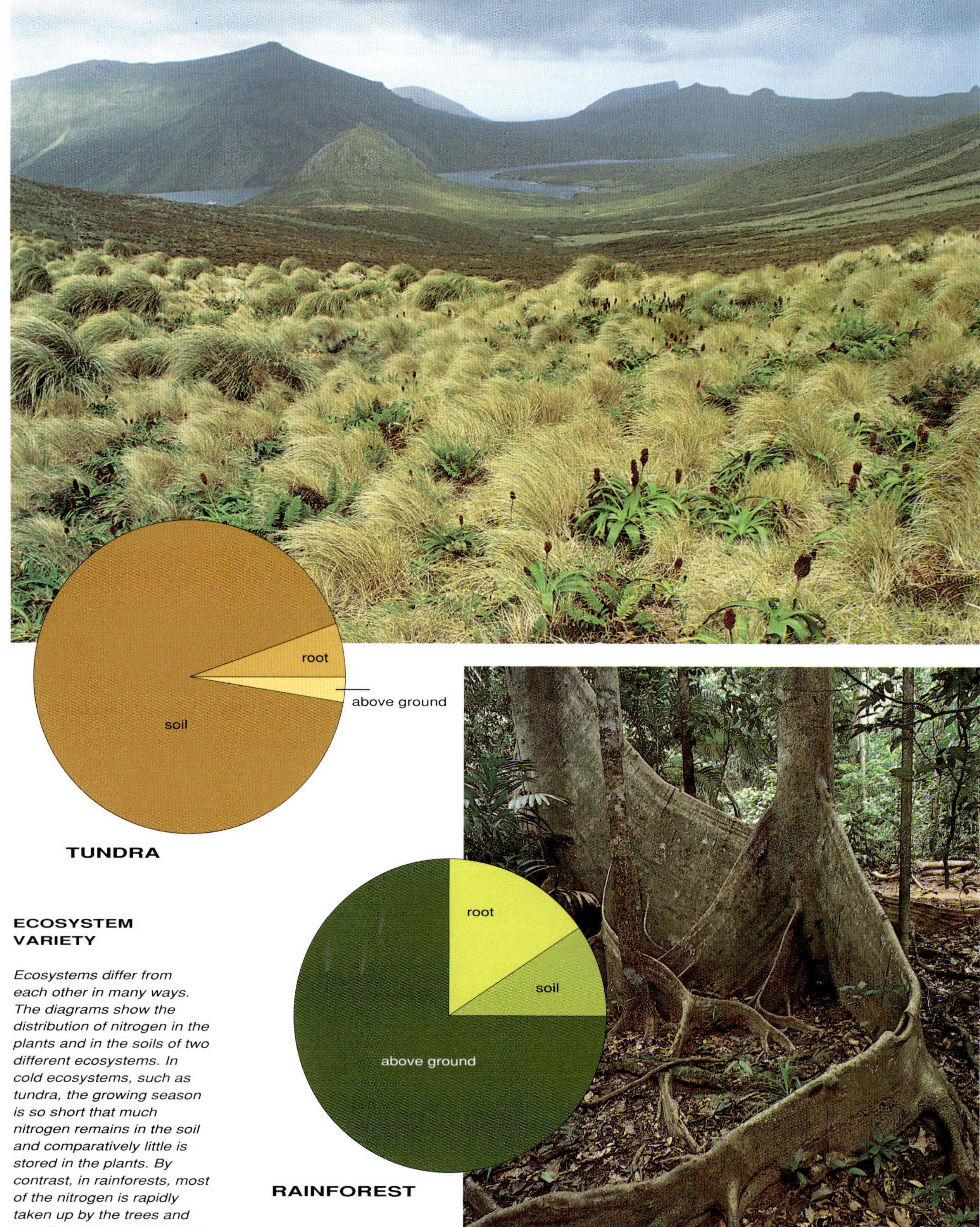

ECOSYSTEM VARIETY

Ecosystems differ from each other in many ways. The diagrams show the distribution of nitrogen in the plants and in the soils of two different ecosystems. In cold ecosystems, such as tundra, the growing season is so short that much nitrogen remains in the soil and comparatively little is stored in the plants. By contrast, in rainforests, most of the nitrogen is rapidly taken up by the trees and very little is found in the soil.

intestinal worms. Plants also have parasites, such as aphids and fungi. Finally, when they all eventually die, their bodies enter relationships with a variety of organisms collectively known as decomposers. These include insect larvae that consume parts of the corpses, and bacteria and fungi that further break them down into their component chemicals, which then become available for recycling.

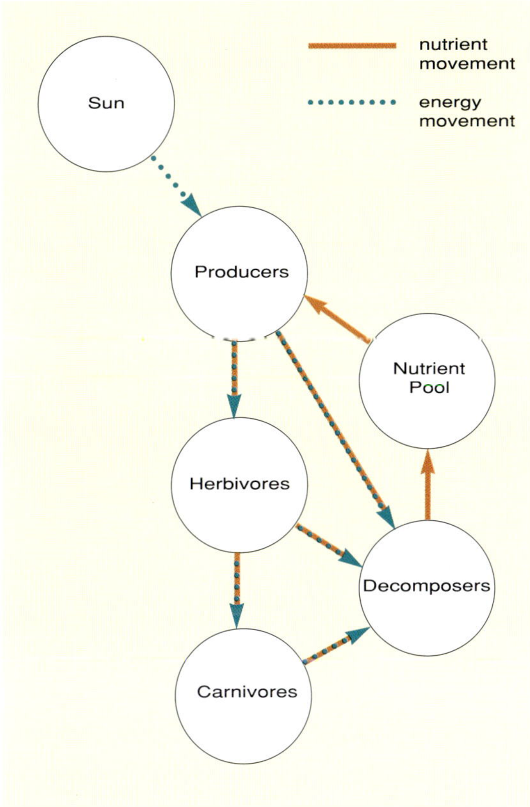

A highly simplified diagram of an ecosystem.

In virtually all ecosystems the primary producers are the plants, which use solar energy to combine carbon dioxide and water to form energy-rich molecules, such as sugars and starches. These molecules, combined with nutrients from the soil, are built up to create cells, tissues and bodies. The energy and nutrients are then passed through the entire ecosystem via herbivores, carnivores, parasites and decomposers.

Ecological processes are basically the same everywhere. However, communities of organisms respond to each particular environment by developing their own variations. This is how the diversity of ecosystems develops. For example, the distribution of the vital nutrient nitrogen varies in different kinds of environments. In very cold climates, nitrogen accumulates in the soil and only a tiny proportion is used by plants because of the limited growing season. By contrast, plants in grasslands and forests consume vast quantities of nitrogen. Consequently these ecosystems support much larger communities of plant-eating animals, which take advantage of the luxuriant plant growth.

There are very practical consequences of understanding such nutrient distributions. In rainforests, for example, most nitrogen is stored above-ground in the trunks, branches and leaves of trees. Clear-cutting removes most nitrogen from the ecosystem, sometimes preventing the forest from growing back. In extreme cases, where inappropriate land management practices follow clear-cutting, areas that once supported dense forests now resemble deserts.

At all levels—species, genes and ecosystems—biodiversity has practical as well as aesthetic value. A natural ecosystem maintains a diversity of species that have evolved together. This diversity of species forms a valuable resource for human welfare; rarely does a year pass without new products being identified and isolated from apparently insignificant organisms. Similarly, the genetic diversity within a species not only provides the raw material for the future adaptation and evolution of that species but also a 'genetic bank' of hereditary material that can be used through the technologies of genetic engineering.

Suggested Reading

Groombridge, B., 1992.
Global biodiversity
Chapman & Hall: Melbourne.

Margulis, L. and Schwartz, K.V., 1988.
The five kingdoms
W.H. Freeman and Company: New York.

Wilson, E.O., 1988.
Biodiversity
National Academy Press: Washington D.C.

Wilson, E.O., 1992.
The diversity of life
Belknap Press: Cambridge, Massachusetts.

Chapter 2

The Origins of Biodiversity

On the time scale of a human life, natural ecosystems appear to be permanent and stable. Unless disturbed by human activities or natural catastrophes, like fire, flood or volcanic eruptions, the grasslands and deserts, forests and tundra appear unchanging. On the time scale of the life of the planet, however, change is the very essence of biodiversity. This chapter looks at the dynamic processes that have created—and are still creating—the profusion of life forms on Earth.

The Beginnings of Life

The Earth originated as a molten ball about four-and-a-half billion (4,500,000,000) years ago. As it cooled, a solid crust formed and steam from the atmosphere condensed in massive thunderstorms to become the oceans. The atmosphere itself was very different from that of today as it contained virtually no oxygen and included many gases that would be lethal to modern life forms. Under these alien conditions, many types of organic molecules—the

LEFT *Microscopic organisms such as these diatoms are abundant in oceans. Diatoms photosynthesise and are just as important as tropical rainforests in removing carbon dioxide from the atmosphere.* **ABOVE** *Fossil animals such as this* Dickinsonia *found at Ediacara in the Flinders Ranges thrived in an ancient ocean 630 million years ago.*

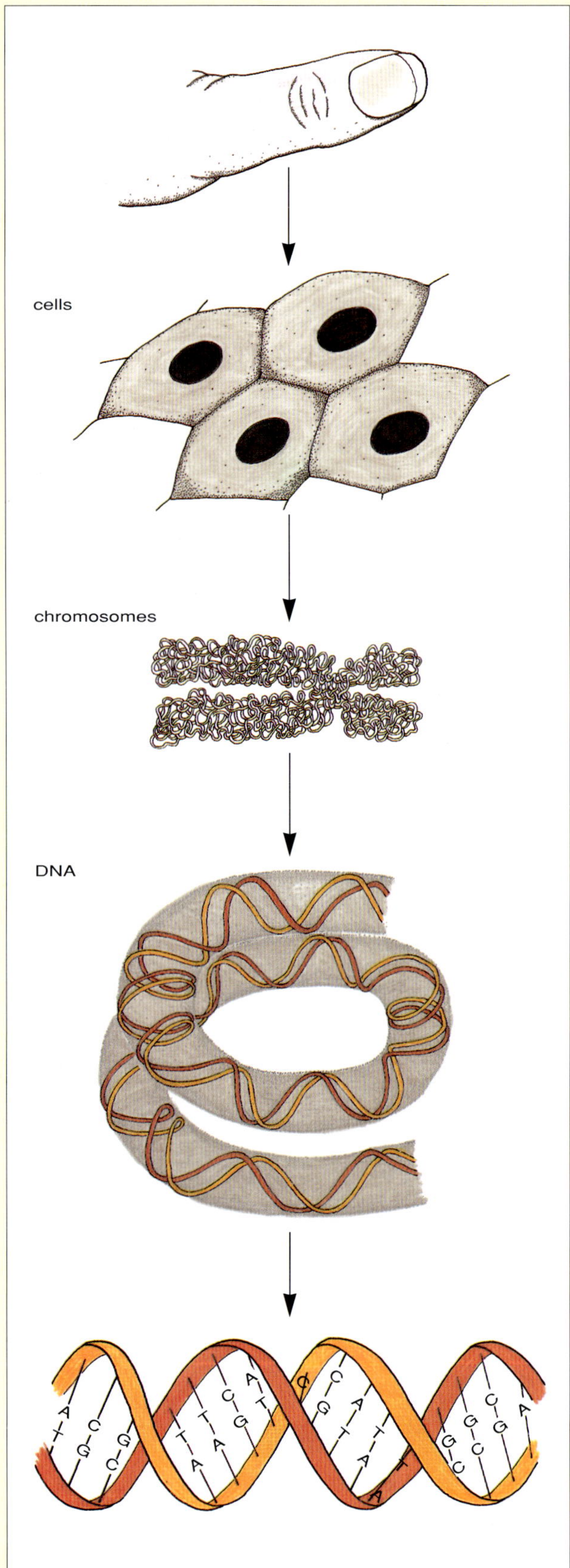

What Information Does DNA Contain?

Genetic material, DNA, is the most sophisticated and compact information-storage system in existence. It is so far advanced it makes modern computer technology look dinosaurian. Just as an example, it takes the fusion of an egg and a sperm to create a new human—but each of these contains only three millionths of a millionth of a gram of DNA. This tiny speck of DNA contains all the information necessary to program the development and function of a totally new person. The current world population of around five billion unique individuals needs only 30 milligrams of DNA—about the weight of a match head—to pass on its genetic inheritance to a new generation of unique people.

Magnified over one million times, the DNA molecule looks like an incredibly long ladder that has been twisted to form a spiral staircase. The uprights of the ladder are made of sugar molecules and the rungs are formed by flat, ring-shaped chemicals called bases. There are four kinds of bases: adenine, guanine, cytosine and thymine, the names of which are abbreviated to A, G, C and T. The sequence of bases in the DNA molecule encodes the genetic information which is read off in threes—rather like morse code. In the latter •••---••• encodes SOS, the international distress call. In DNA the sequence TACAATAGTGGGCGG translates to 'start here' (TAC) then 'join together the amino acids leucine (AAT), serine (AGT), proline (GGG), alanine (CGG)' to form a specific protein. The direct product of genes, proteins are the building blocks of all cells.

LEFT *DNA molecules are shaped like twisted ladders, where each 'rung' is a pair of chemicals called nitrogenous bases. The pairs are specific so that rungs are made of either adenine (A) and thymine (T) or guanine (G) and cytosine (C). The genetic information is coded into the sequence of bases when read along one side of the ladder. The DNA is coiled into microscopic structures called chromosomes. Each human cell, such as one taken from a thumb, contains nearly two metres of DNA, neatly packaged into 46 chromosomes.*

The stromatolites at Shark Bay in Western Australia are constructed by micro-organisms called Cyanobacteria. Fossil stromatolites from Western Australia are 3.5 billion years old and are among the earliest evidence of life on Earth.

building blocks of life—accumulated in the oceans through simple chemical reactions. A natural property of organic molecules is their tendency to join together to make more complex molecules and for these, in turn, to aggregate into larger, but still microscopic, structures. Some of these structures resemble simple cells. As the number and variety of these forms built up in the primeval oceans, the supply of organic molecules essential for their energy needs and structure became depleted. Hence the first struggle for survival took place. Some forms developed the ability to harness solar energy and to use this to manufacture their own energy-rich molecules and building blocks. This process, which we know as photosynthesis, originated in forms that resemble Cyanobacteria (often called blue–green algae). Masses of Cyanobacteria forming dome-shaped structures called stromatolites can be seen at Shark Bay in Western Australia. Fossil stromatolites from this State date back an astonishing three-and-a-half billion years, representing some of the earliest evidence for life on Earth.

A by-product of photosynthesis, the release of oxygen, radically changed the planet. As oxygen accumulated in the atmosphere, most of the prototypes of life, which had developed in an oxygen-free world, died out. Those few that survived became the ancestors of our current life forms and present-day biodiversity.

Genes and Mutations

An essential step in the evolution of truly living things from the chemical 'blobs' in the ancient oceans was the development of an ability to transfer information from parents to offspring. Instructions for building and operating a complex body are passed from generation to generation as coded chemical messages called genes. This hereditary mechanism ensures that offspring resemble their parents and, through interbreeding outside their immediate families, maintains the genetic variation of each species. In flies it takes about 10,000 separate genes to code the instructions for eye colour, number of legs, body shape, behaviour and all other attributes that make up each individual. Humans have more complex bodies and functions, which require about 100,000 genes to code all their attributes.

The specifications for a hereditary information system are extremely strict. It must be very robust such that instructions do not become garbled over the generations. At the same time, it must be capable of limited change so that structure and function can slowly evolve in response to changing environmental conditions. Finally, the system must be amazingly compact so that tens or even hundreds of thousands of instructions can be packaged into

Mutations—The Good, the Bad and the Neutral

Most mutations are detrimental to the individuals inheriting them. This makes sense as the genes common in a species at any one time are those that have been favoured by many previous generations of natural selection. Any mutational change in a gene is, therefore, most likely to impair rather than improve its function. However, some mutations may alter the gene so slightly that they do not really affect function—neutral mutations. Rarest of all are mutations that increase the chances of survival of the individuals inheriting them—beneficial mutations.

An excellent example comes from over 150 years of observation of the Peppered Moth in Britain. In the early 19th Century, virtually all of these insects had mottled wings camouflaging them on lichen-covered trees, thereby protecting them from predatory birds. There were minor variations in the amount of dark versus pale in their wing patterns but these neutral variations did not alter the effectiveness of their camouflage. Occasional detrimental mutations produced all-black moths: easy targets for hungry birds. However, the onset of the industrial revolution reversed the action of natural selection. Sulphurous fumes destroyed the lichens and blackened the tree trunks, so the original mottled moths were no longer camouflaged and became easy picking for birds. By 1950, in some areas, nearly 100 per cent of the moths were of the black mutant type. A mutation that had been deleterious in a pristine environment had become beneficial in a polluted one. However, evolution is still at work: clean air regulations in Britain are resulting in a slow return of lichen in industrialised areas, once more favouring the peppered form of moth.

the microscopic sex-cells, which are the only link between one generation and the next. It appears that only one mechanism has met these specifications—at least on this planet. All organisms rely on the same mechanism, the information being coded into the chemical structure of the long, thin molecules of DNA.

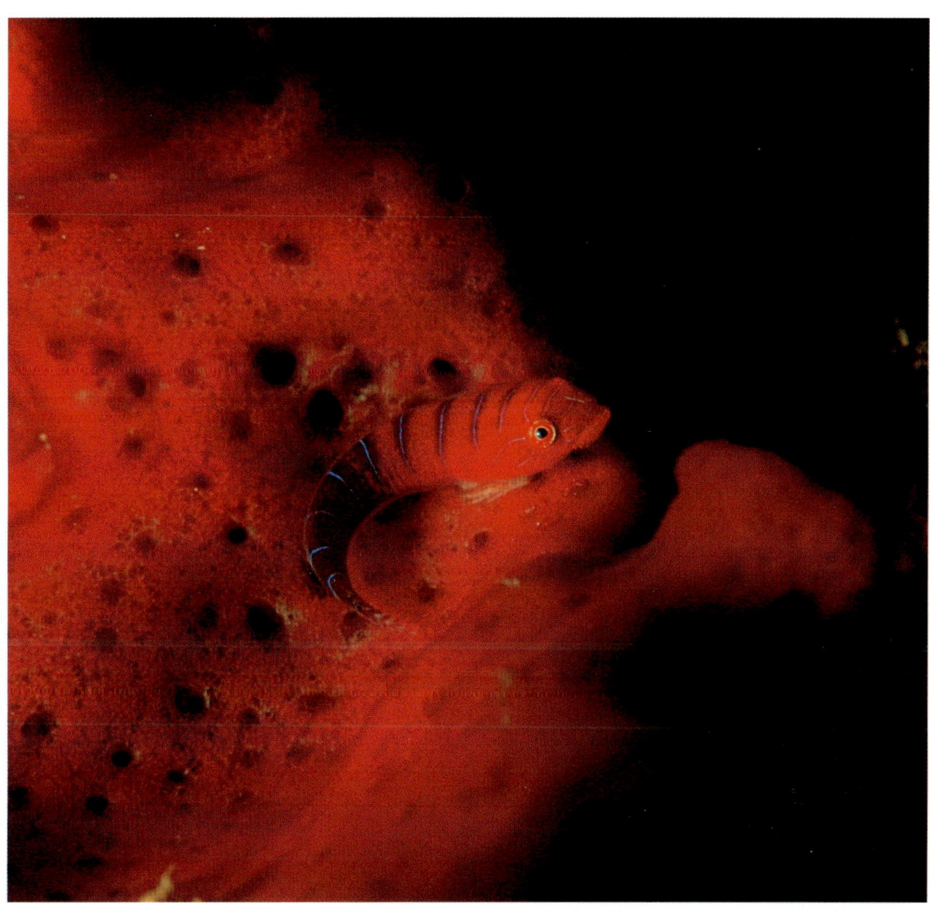

Within each cell, the DNA is coiled into structures called chromosomes. For example, in our bodies every cell contains 1.8 metres of DNA molecules packaged into 46 chromosomes. There may be several thousand genes along a single chromosome. The information in each gene is decoded by complicated biochemical processes in the cell to produce all the instructions and molecules required to create a living organism.

If all living things on Earth contained identical DNA then there would be only one species of organism and every individual would be an identical twin of every other. Life would be incapable of evolving. However, while DNA is generally transmitted from generation to generation with great fidelity, occasional mistakes do occur. These are called mutations and are the ultimate origin of all biodiversity.

Mutations are generally tiny errors in the DNA. They can arise from miscopying, when the DNA is replicated within a cell, or from chemical or radiation damage. Mutation can appear randomly anywhere in the DNA. Obviously, if mutations accumulated unchecked in the DNA, the informa

These red and yellow clingfish are actually the same individual fish, which changes colour to blend in with the background. This kind of adaptation is used by many animals to avoid predators but it can also make it difficult to identify different species.

tion would become meaningless and, after many generations, would not be capable of coding for a functional organism. This would be analogous to the General at the battlefront who sends the message 'Send reinforcements, we're going to advance'. On reaching headquarters, after passing from soldier to soldier, the message becomes 'Send three and fourpence, we're going to a dance'—not a valuable instruction! In nature, the information in the DNA is continuously edited by the process of natural selection.

Natural Selection

Over 130 years ago Charles Darwin proposed the concept of evolution through the mechanism of natural selection. Since then, our expanding knowledge of heredity has been integrated into this theory and we can now understand how mutations in the genetic message are edited.

Most organisms appear to produce far more offspring than are necessary to establish the next generation. However, the resources available to any species are limited (for example, nest sites, suitable food or soil), and, on average, only two offspring will survive and reproduce in their turn. Competition among offspring is therefore inevitable. In this competition, the particular combination of genes each individual has inherited is critical to its chance of success.

The least fortunate individuals are those inheriting one or more mutations so defective that the embryonic organism dies (lethal mutations). Others may receive genetic combinations that impair their structure or function such that they are unlikely to succeed in the struggle to find food, homes and mates (detrimental mutations). Other mutations are so subtle that they really do not influence the survival chances of the individual inheriting them (neutral mutations). The rarest types are beneficial mutations, which endow their recipients with features that increase their chance of surviving and reproducing. Over succeeding generations, these mutations will spread throughout the species, displacing the original version of the gene and bringing about an evolutionary change. This is the process of natural selection.

It can be seen that the speed of genetic editing depends upon what effect a mutation has. Lethal and detrimental mutations are quickly removed from the genetic repertoire of a species. Neutral mutations may persist for longer—just by chance—while beneficial mutations become a feature of the genetic makeup of the species.

Genetic Variation and Adaptation

Most species are distributed patchily over their ranges and are subdivided into local populations. The scale varies enormously: populations of fungi may occupy different types of logs in a forest; populations of butterflies, different patches of forest; and populations of dolphins, different oceans. As a result, each population encounters a different environment and in each environment a slightly different combination of genetic mutations may be favoured by natural selection. For example, a slender body form may be beneficial to a herb competing for light with long grasses in a pasture. By contrast, in individuals of the same herb species growing on a windswept cliff-top, mutations for a prostrate growth form would be beneficial. The genetic fine-tuning of populations to suit them to their local environments is called adaptation, and this maintains the wealth of genetic variability within each species.

Speciation and Extinctions

The evolutionary process does not only bring about changes within a species over time, it also generates a diversity of new species. Imagine an ancestral species distributed over a broad area. When geological changes occur, such as islands cut off by a rise in sea level, volcanic lava flows across the land or the desertification of central Australia, populations of this species may become totally isolated from each other. In isolation, the populations will undergo independent adaptation and, over time, can become genetically distinct. If these populations were to encounter each other after further environmental changes they may have changed so much that they do not recognise each other as potential mates or, if they did mate, the resulting hybrid offspring might be inviable or sterile. As a species is, by definition, a group of interbreeding individuals, the ancestral species has become two new species. The process of producing new species is called speciation.

Cliff-dwelling plants adapt by growing close to the rock face, reducing the chance of being blown away.

The evolution of the Australian whipbirds is a good example. Changes in climate isolated populations of the once-widespread ancestral species. The whipbirds in south-western Australia became adapted to drier conditions in heathland and mallee scrub, while those in the east adapted to rainforest and wet sclerophyll forest. The Western and Eastern Whipbirds now differ in preferred habitat, plumage and song and are two distinct species.

The fossil record shows that hundreds of millions of species have evolved since life first began on Earth. But the planet has not simply continued to fill up with more and more species. Instead, the process of speciation is roughly balanced by an opposing process—that of extinction (the death of all members of a species). In fact, palaeontologists (scientists specialising in fossils) estimate that over 99 per cent of all species that ever existed on Earth have become extinct during the immensely long time since the origin of life.

Extinction takes two forms. The first involves environmental catastrophes—swift changes in global

Different Ways of Looking at Biodiversity

The number of species occupying a particular habitat is known as the species richness of that habitat. This number may also be called the alpha diversity *of the habitat.*

Beta diversity *is a measure of the similarity (or difference) of a series of habitats. Where there is little similarity in the species occurring in different but adjacent habitats, high biodiversity can be expected. Conservationists are often interested in areas of high beta diversity because there are many different kinds of habitats each occupied by different species. (However, it is important to note here that these individual habitats do not require a high alpha diversity to be important to conservation if the few species they harbour are rare or simply not found in many other places.)*

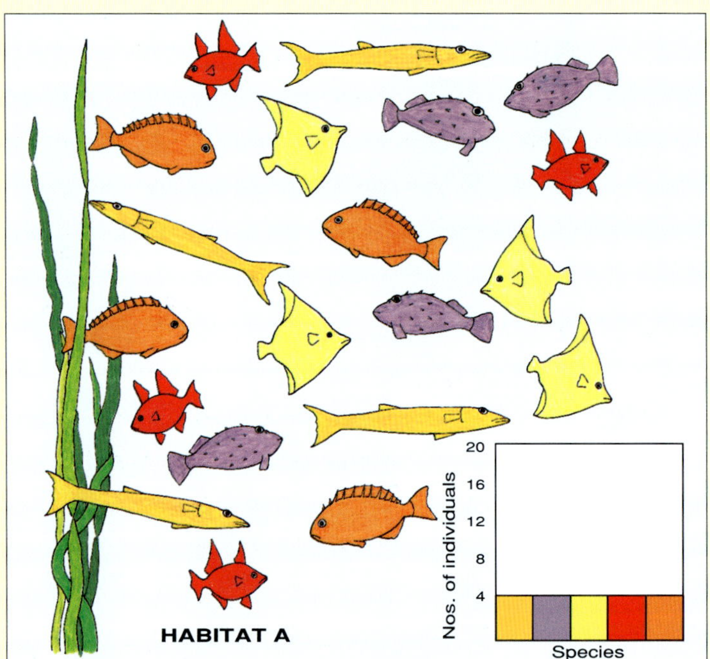

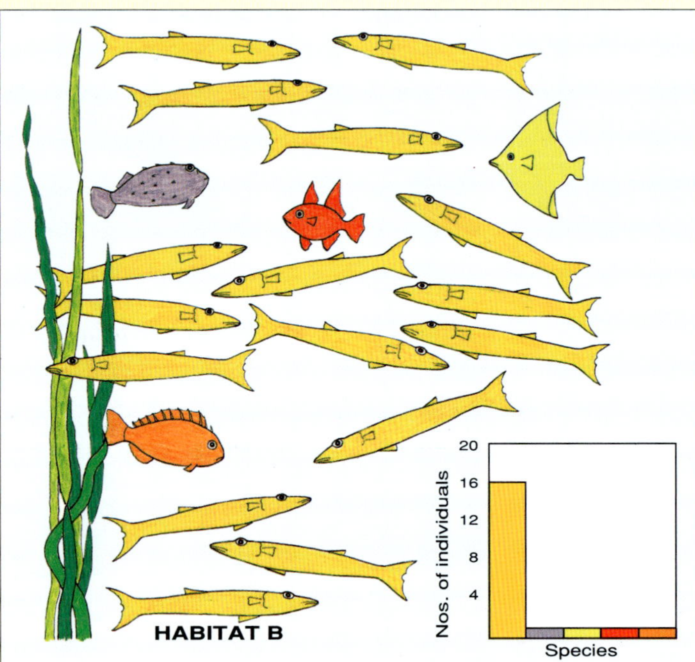

Endemism *refers to those species in a community or habitat that occur nowhere else. Endemism tends to be high on islands where evolutionary processes have been isolated from the rest of the world. Deep cave systems harbour many endemic species for the same reason. Australia, being essentially a huge island, has many endemic species (see Chapter 7).*

Evenness or equitability *is a very important component of biodiversity because it takes into account the commonness or rarity of species. The figures here show two habitats, A and B, each with identical species richness: 5 species each. However, in habitat A, every species is equally common, whereas in habitat B, one species is very common and the others are rare. Measures of evenness show that habitat A is more diverse than habitat B because in habitat A it will be relatively difficult to tell which species you will encounter next as they are all equally likely to cross your path. By contrast, in habitat B, the next species to be encountered is easier to predict as it is very likely to be the common one. This example involves just a few species so that it is readily understood. The real value of the evenness measure is best appreciated when comparing habitats containing hundreds or thousands of species.*

climate, massive asteroids hitting the Earth with the power of thousands of nuclear weapons, or the collision of continents. Such catastrophes have caused many mass extinctions, such as the demise of the dinosaurs. Often the mass extinctions were followed by periods of speciation covering only a few million years (an immense time from our perspective but only the blink of an eye on a geological time scale) in which hundreds of thousands of new species arose. The second reason for extinction is more subtle. Each species faces challenges from its surroundings, both from climatic and other physical factors as well as from other species that compete with it for food and space or attempt to eat it. Over a period of time, the ecological circumstances of a given species may slowly deteriorate until it can no longer exist. This perpetual process causes a slow background extinction rate—a steady loss of species in addition to the mass extinctions.

What Forces Determine Diversity?

The total amount of biodiversity on the planet at any time is determined by the opposing forces of speciation and extinction. However, diversity at any particular site is influenced by four key factors: climate, colonisation (the ability of organisms to arrive and settle at the site), time and variety of habitats (how many different opportunities the site contains for species to exploit).

Climatic Conditions

The global circulation of air and ocean currents is the major determinant of climate and, therefore, biodiversity. There are major gradients of increasing levels of light, temperature and longer growing seasons from the poles to the equator. These gradients result in conditions more favourable to plant growth closer to the equator. Similar patterns can be seen in the richness of vegetation from high altitudes to sea level. As a consequence, biodiversity tends to be highest at low latitudes in equatorial regions—for example, in tropical lowland rainforests—and lowest near the poles—for example, the vegetation within the Arctic and Antarctic circles. However, peculiarities of air and ocean circulation and major geographical features, such as mountain ranges or islands, disrupt this apparently simple pattern to produce important regional variations—for example, the remarkably high levels of biodiversity in Mediterranean climatic regions (see Chapter 6). Also, within this broad framework, many subtle processes associated with local microclimate occur. In tropical forests, for example, the light hitting the canopy is very intense but in the shadows beneath there is a gradient of light intensity towards the ground. This gradient provides a variety of habitats, each defined by different light spectra, to which different species of plants may become adapted.

Another important factor is seasonality. Some annual plants, for example, are adapted to different seasons and grow at different times of the year so that many species coexist in the same place.

Colonisation

New species arise in relatively small geographic regions. Their ability to expand into other areas, and therefore to add to the diversity of those areas, depends on their biology and on the nature of the physical barriers that separate the place of origin from other areas. Marine organisms may encounter relatively few physical barriers—for example, juvenile Great Barrier Reef organisms are carried southwards thousands of kilometres by ocean currents. Seeds and animals such as lizards and rodents can cross the oceans on rafts of floating vegetation. Similarly, spores of micro-organisms, winged or tiny seeds and small animals may travel passively on air currents, while flying insects, birds and bats can actively explore great distances.

In contrast, large terrestrial animals, which may be capable of migrating over considerable distances, can be barred by relatively narrow water

Cradle Mountain National Park in Tasmania is an ecosystem that contains several natural communities, including the shrub community and the communities of lichens on the rocks.

ways. Many terrestrial invertebrates have such exacting habitat requirements that only a few kilometres of inhospitable territory may effectively prevent their expansion.

The fate of immigrant organisms is determined by the features of their destination. Young Great Barrier Reef fish do not establish on the south-eastern Australian coast because water temperatures are too low. In the same way, seeds need appropriate soils and climate, and animals, appropriate food sources.

The processes of colonisation are not just important in understanding the history of current biodiversity. They are also essential in the planning of reserves and rehabilitation programs. Most rehabilitation programs focus first on establishing the plants with the expectation that animals will then recolonise.

Time

The biodiversity at any specific site also reflects a time element—the history of the site. This is well illustrated by Krakatau in the Java Sea. This island exploded during a volcanic eruption in 1883, leaving a lifeless fragment about five kilometres long. When the smoke cleared, the biodiversity of this fragment was zero but, through time, many kinds of organisms found their way there—mostly on air currents. They included fungi, small plants and a great variety of invertebrates. In recent years, grasses, ferns and trees have colonised the island, as have a variety of birds and other vertebrates. Clearly this process takes time, so older, geologically stable areas tend to harbour more species than younger ones.

The island concept can be applied broadly in

the context of colonisation. For example, when a plant species successfully invades an area it has not previously inhabited, it may find itself surrounded by a 'sea' of other species. This 'sea' may act as a barrier to the specialist consumers of these plants, which may take a long time to find it in its new area. This process sometimes explains why some plant species are attacked by many kinds of herbivores, while others have only a few or none at all. It also explains why plants and other organisms brought to new environments often get out of control and become pests. Privet, for example, grows abnormally well in Australia because it has been brought here without its natural control agents. If we do not intervene, it may take centuries for its herbivores to migrate here from overseas, or thousands of years for local herbivores to evolve the ability to control privet populations.

The Variety of Habitats

It is useful to think of the variety of habitats as operating at three different scales. First, landscapes that are mosaics of geological features, such as deep valleys and high mountains or wide rivers and arid areas, are likely to harbour more species than uniform terrain. This is because there are more different environments available. Second, within a particular kind of environment—for example, an area of grassland—patchworks of different features such as rocky outcrops, termite mounds and clumps of bushes create smaller habitats to which species may adapt. More species can become established in a varied area than a uniform one. Third, in highly varied environments with many smaller habitats, there appears to be a positive feedback in which the wealth of species that take advantage of the wide variety of habitats in the first place provide, in turn, habitats for still more species. Take, for example, a forest with a mosaic of tree falls, rotten logs and different kinds of soils. The plants specialising on these often harbour their own array of specialist insect herbivores, pollinators and seed dispersal agents, plus their parasites and predators. The habitats of many of these species may be tiny—often restricted to one part of one plant species. When space is divided up in this way, many habitats and their many specialist species can be 'packed' into small areas.

Human Impact on Evolution

The expansion of our species has initiated a major episode of mass extinction. Habitat destruction, hunting, the transfer of diseases, pest plants and animals all have a severe impact on biodiversity. For example, foxes can decimate populations of Australian native mammals. Reduction in population size decreases the genetic variability within a species and, therefore, the ability of that species to adapt to a future change in the environment (for example, global warming). Moreover, extinctions caused by human activities are now occurring in the time frame of centuries or even decades, rather than millions of years. This leaves no time for adaptation and speciation to replace the species we are losing.

Loss of biodiversity has a very important practical dimension, as well as aesthetic and ethical considerations. Each living species can be viewed as an adaptive triumph over a long series of challenges—simply because it has survived. Coded in the DNA that makes up the genes of each species are the solutions to problems past and present. As we share many challenges with other species—countering attacks by disease organisms is just one example—their adaptations may be our solutions.

Suggested Reading

Archer, M., Hand, S.J. and Godthelp, H., 1991.
Riversleigh: the story of animals in ancient rainforests in inland Australia.
Reed Books: Sydney.

Australian Academy of Science, 1990.
Biology: the common threads.
Australian Academy of Sciences: Canberra.

Cockburn, A., 1991.
An introduction to evolutionary ecology.
Blackwell Scientific Publications: Melbourne.

Chapter 3

Biodiversity and Bioresources

Biodiversity is the biological wealth of this planet and is fundamental to every culture and civilisation. It contributes directly to the world economy in many, largely unseen, ways. For example, we are all aware of the economic benefits provided by plant and animal crop species. We are often less aware of the potential stored in their genes, or those of their wild relatives, that may be of critical value in responding to future market and climatic demands.

Biodiversity is often difficult to value financially because we still do not know all of the parts that may be useful to us. However, some products of biodiversity can be valued in very specific ways. While it might seem odd to discuss life in dollar terms, it is useful to look at what micro-organisms, plants and animals do for the economy. This chapter will show that biological resources include not just species but also genes and entire ecosystems.

Green ants are used by traditional Aborigines to treat respiratory disorders. Another species of weaver ant, being a fierce predator, has been used for centuries in China to control orchard pests.

Conventional Bioresources

These are the familiar plants and animals that humans have used—some for thousands of years: cattle, pigs, sheep, chickens, goats, horses, wheat, barley, oats, rice, corn, honeybees, molluscs, crustaceans and fish for food; many species of trees for paper products, furniture and building materials; a myriad of plants for medicines, fibres, dyes, pesticides, flavourings, oils and clothing, and many micro-organisms, including fungi and yeasts for bread-making, brewing and dairy products.

When all human cultures, past and present, are considered, they exhibit an astonishing array of dietary and cooking habits, and the use of numerous species for food. The human digestive system is so versatile that few areas of biodiversity are excluded from our diet.

Humans have always been resourceful hunters and gatherers using many kinds of land, freshwater and marine creatures (like insects, molluscs and fish) as well as fungi, and a great variety of plants, including the roots, leaves, flowers, seeds and fruits. Basic technological advances like the invention of stone tools made it possible to cut up fresh meat, smash bones to extract the marrow, dig succu lent roots and tubers from the ground, and skin animals thus increasing the array of useful species. Human mastery of fire meant that woody plants took on a new importance, becoming a useful resource. Such advances during the course of human evolution have resulted in our ability to exploit biological resources that were previously of little value. As humans spread to different parts of the world, the new animals and plants they encountered were used, taking advantage of an ever-increasing proportion of biodiversity.

Aboriginal Bioresources

Aboriginal people have developed complex relationships with the environment, seeing themselves as integral parts of the landscape. Traditional Aboriginal diets are varied, with a far greater diversity of plant and animal foods than those consumed by many contemporary societies. In south-eastern Australia, much knowledge of wild foods has been lost but, in the north and west, many Aboriginal people continue to supplement their diets with a wide variety of foods obtained by hunting and gathering.

Bush medicine is also an important component of traditional Aboriginal life. Many compounds produced by plants are recognised and used to treat a wide range of ailments. Aromatic herbs, tannin-rich inner barks and gums have well-documented therapeutic effects. European medicine has been slow to recognise the value of indigenous medicines.

We can learn much about the uses of biodiversity from people who have been studying it for thousands of years. At Uluru in the Northern Territory, ecologists are working with the traditional owners to better understand the impact of both exotic species and increased tourism on biodiversity.

The Value of Wild Genes to Agriculture

Humans have been engaged in a war against crop pests—especially viruses, bacteria, fungi and insects—since cultivation first began. Even in countries with high-tech farming, the widespread planting of a single, high-yield crop variety (to maximise profits) makes agriculture very vulnerable to pests. This is because diseases and pests which can overcome the resistance of that particular variety are presented with the opportunity to attack entire crops.

One vital defence against crop and livestock disease involves finding wild relatives of the crop

Many products of native plants are now being cultivated to supplement our diet as well as farm income. Increasing the diversity of food resources could give us security against tough times, something that reliance on a handful of crops does not always provide. Native plants shown in this dish are as follows (clockwise, from top left-hand corner): lemon myrtle leaves, riberries, Warrigal greens, macadamia nuts, emu eggs, dried bush tomatoes, mountain pepper leaves, boab nut, bunya nuts, fresh bush tomatoes and kurrajong seeds.

Many industries rely on biodiversity, in particular tourism. On Heron Island, a channel was cut through the reef to allow boat access for tourists. When such changes are made it is vital that biodiversity is maintained to continue attracting tourists.

plants and domesticated animals that are resistant and then breeding this resistance into domesticated varieties. As a result, wild relatives are particularly important components of biodiversity. In recent years, many countries have collaborated to send expeditions to the original sources to collect living specimens, either for breeding programs or for storage in gene banks. These expeditions have gathered many useful genes, including those that confer resistance, and others that enable crop plants to grow with less fertiliser. The wheat, maize, sorghum, barley, potato, soybean, rice, grape and cotton industries have each profited from wild genes both in terms of yields and dollars. Sometimes, the profits have been staggering. For example, the United States Department of Agriculture estimated that for 1980 alone, wild genes increased agricultural profits by $US1 billion.

The advent of agriculture and the domestication of plants and animals enabled most modern human societies to provide themselves with reliable sources of food. However, two long-term effects have caused concern. First, our reliance on domesticated animals and plants has, in general, drastically reduced the variety of foods in the average diet. Most cultures rely on only a few crop and livestock species and news reports frequently show malnutrition and starvation resulting from the failure of just one. The vast majority of the population of the world relies almost entirely on a handful of crops: wheat, maize, potatoes and rice. Also, demand for the most commercially viable crops, such as those with the highest yield, has led to concentrated production of a few varieties. Consequently, the genetic diversity of crop plants has become significantly reduced. Second, clearing vegetation for pastures and crops has had an enormous impact in reducing the biodiversity of the world. Vast forests have been cleared from the temperate zone for crop plants or grazing animals, reducing the remaining biodiversity merely to the soil organisms. While these changes were necessary, the cost in terms of the reduction of biodiversity has been immense.

Biodiversity-based Industries

Mining exploits the mineral wealth of Australia but the other pillars of the Australian economy—agriculture, pastoralism, forestry, fisheries and tourism—depend upon our *biological wealth:* biodiversity.

Rural industries rely on biodiversity to maintain the health of soils in which crops, trees and forage plants grow. Many pastoralists use the diversity of wild rangeland plants to provide forage for domestic stock. Timber companies utilise the existing diversity of forest trees. Many crop, forage and timber species, in turn, depend upon biodiversity for pollination, defence against herbivores and other pests, and for the dispersal of their seeds and fruits. For example, the roots of many timber species require specialised fungi to absorb nutrients from the soil; predators such as spiders and carnivorous beetles suppress pest populations in crops; wild native bees provide pollination for tropical orchards; and the fruits and seeds of trees and forage plants are dispersed by animals as diverse as small marsupials, bats, birds and ants.

The fishing industry harvests wild species from the biodiversity of freshwater and marine ecosystems. Each commercially important species in turn feeds on other components of biodiversity, and many rely on the biodiversity of mangrove, estuarine or reef areas for successful reproduction. For example, juvenile fish hide from predators among seaweed; invertebrates are the food of many of the fish we eat; and game fish are predators of other fish. Shellfish feed by filtering bacteria, single-celled animals and plants (plankton) and small invertebrates from the water. When all these components are considered together—bacteria, invertebrates, plankton and seaweed—it is clear that the fishing industry uses many components of biodiversity, either directly or indirectly.

Tourism is a growing biodiversity-based industry. Visitors from across Australia and overseas come to see our forests, deserts and reefs for their natural beauty and for the animals and plants that occur in them. Our most charismatic attractions such as the Koala, kangaroos, brilliantly coloured reef fishes and whales, all depend, in turn, on biodiversity for their food.

In addition, there are many parts of biodiversity that Australians may take for granted but overseas visitors find particularly attractive. These include the Emu, the Jabiru, gum trees, wattles, orchids, parrots, kookaburras, blue-tongued lizards, goannas, fruit bats, bird-winged butterflies, echidnas, crocodiles and a host of marine life, such as corals, sea urchins and even multi-coloured sea slugs, which may not be found in the home country of the visitor.

The protection of biodiversity is essential for tourism as visitors do not come to see dying coral reefs, denuded hillsides, overgrazed arid lands, mud-filled rivers or poorly planned developments scarring magnificent forest or desert landscapes.

Biodiversity is therefore the basic resource for tourism, agriculture, pastoralism, forestry and fisheries. Unfortunately, its value to these industries (trillions of dollars annually worldwide) has never been calculated because it has not yet been factored into conventional accounting nor has it become part of mainstream economics.

Other major industries rely heavily on biodiversity. One example is the pharmaceutical industry, which screens wild plant species for new drugs. Its basic resources are the undisturbed forests of the world, especially in the tropics, which hold hundreds of thousands of unexplored species. Screening programs have had some remarkable successes. The Moreton Bay Chestnut, for example, is a native of the forests of eastern Australia, and contains a chemical that has proved useful in the fight against AIDS. Worldwide sales of

Capturing Energy From the Sun: the Most Basic Ecosystem Service

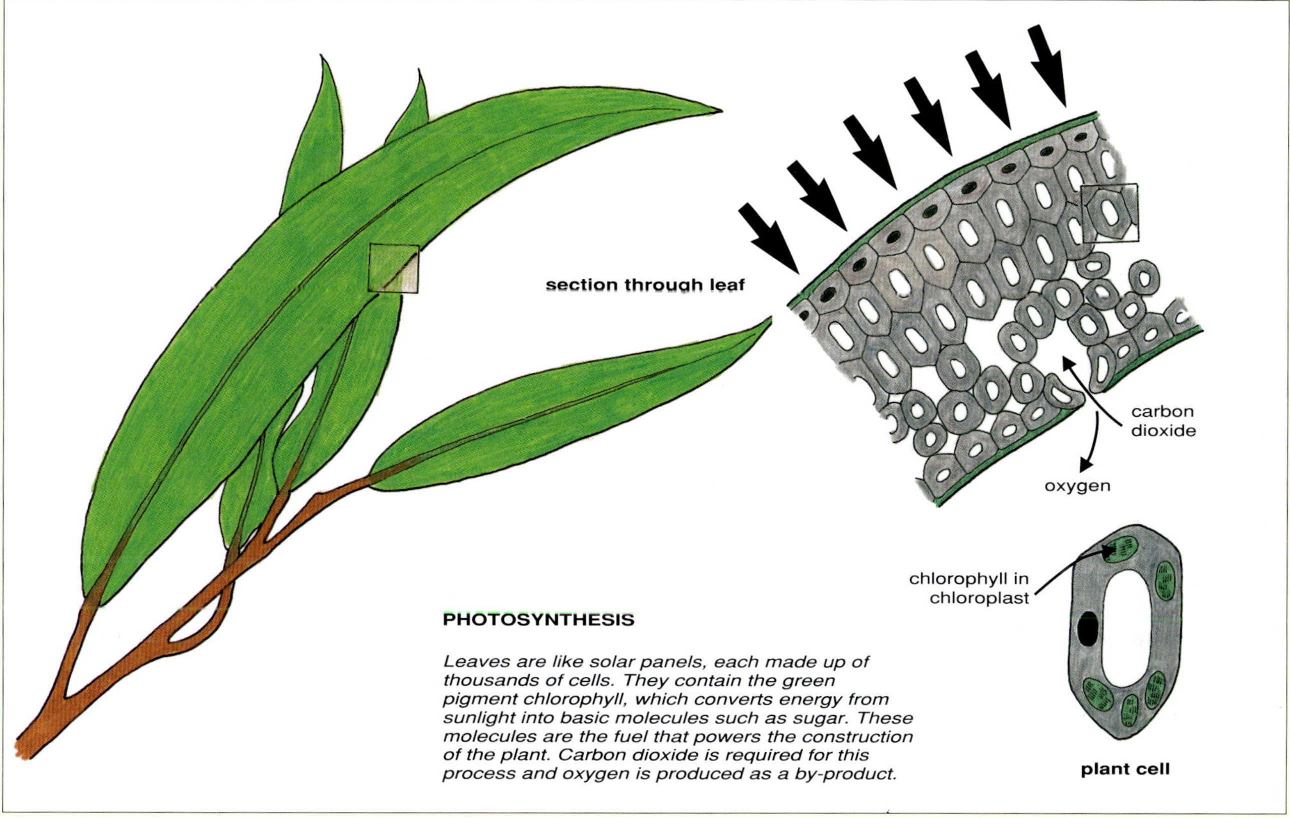

PHOTOSYNTHESIS

Leaves are like solar panels, each made up of thousands of cells. They contain the green pigment chlorophyll, which converts energy from sunlight into basic molecules such as sugar. These molecules are the fuel that powers the construction of the plant. Carbon dioxide is required for this process and oxygen is produced as a by-product.

Plants possess the ability to capture energy from sunlight and convert it into a basic fuel—sugar. This process, called photosynthesis, takes place in cells where a green, light-harvesting chemical called chlorophyll uses solar energy to combine water and carbon dioxide to make sugar. A by-product is oxygen. On land, most photosynthesis is carried out in leaves. In the oceans and fresh water, it is mostly the product of single-celled algae collectively known as phytoplankton.

Billions of years ago the sun bathed a lifeless Earth with its light and heat. The evolution of photosynthesis provided a pipeline that channelled some of this immense solar energy into living organisms, fuelling the food chains and establishing thriving communities.

A serious reduction in the photosynthetic capacity of the Earth would be like reducing the power supply to a factory. Many machines and appliances would stop, the lights would dim and the images on computer screens would shrink to the size of postage stamps. Without photosynthesis, the biodiversity of the Earth would shrivel—only some bacteria and the communities dependant solely on them would remain.

Plant biodiversity is so vast that photosynthesis occurs in most environments. Different plant species are adapted to different conditions: saltbush and spear grass are typical plants in the arid outback, seaweed and plankton in the sea, and gums and wattles in woodlands. One practical application is the use of the adaptations to aridity in selected Australian Eucalyptus *species. These trees are exported to countries to reforest areas denuded by drought.*

Nicola Oram '93.

pharmaceuticals containing active ingredients derived from plants are estimated at $US40 billion a year.

Some industries are extractive, utilising natural biodiversity as their raw materials. Familiar examples include fishing and forestry. However, tourism requires biodiversity intact and undisturbed where it is found. Visitors come to see the beauty of our forests and reefs intact, and to enjoy the myriad of animals and plants these places harbour. The Australian tourism industry earned $18 billion in 1990–91 alone.

Ecosystem Services

The concept of ecosystem services brings together an apparently odd pair of subjects—ecology and economics—but it is important to note that they both come from the same Greek word that means 'to keep your house in order'. Writing in the first issue of the international journal *Ecological Economics* (1989) Professor Paul Ehrlich wrote of ecosystem services: ...*'The majority of economists have never been taught that ecosystems provide humanity with an absolutely indispensable array of services, including maintenance of the quality of the atmosphere, the amelioration of the climate, operation of the hydrological cycle (including flood control and the provision of fresh water to agriculture, industry and homes), the disposal of wastes, the recycling of nutrients essential to agriculture and forestry, the generation of soils, pollination of crops, pest control, the provision of food from the sea, and maintenance of a vast genetic library from which humanity has already withdrawn the very basis of its civilisation. While these services are "free", they would, of course, be infinitely costly to replace.'*

LEFT *An important method of biological control is to increase the biodiversity in crops so that predators, parasites and pathogens can attack pests. In the bottom left corner an antechinus eats a grasshopper, just above, a huntsman spider eats another; immediately to the left the larva of a ladybird beetle combs a stem for greenfly, while on the leaf above a caterpillar erupts with hatching parasitoid wasps whose larvae have consumed its internal organs. Above the dead caterpillar a praying mantis feeds on a fly while next to it an adult ladybird beetle searches for greenfly. To the right, a net-casting spider traps an ant. On the ground below a bull ant carries its prey and a solitary wasp drags a paralysed cicada back to its burrow. The wasp larva that hatches from the egg laid on the cicada will have a supply of fresh meat to feed on. In the right foreground the corpse of a caterpillar killed by an insect-attacking fungus bursts open at one end as the club-shaped fungal fruiting bodies prepare to release their spores to infect further prey.*

Farming is a biodiversity-based industry. Soil biodiversity maintains most soil fertility while natural predators, parasites and pathogens are responsible for much pest control. Manipulation of biodiversity may reduce the heavy dependence on chemical pesticides in crops such as cotton.

The phrase 'ecosystem services' emphasises how human activities rely heavily on ecosystems functioning normally, whether they occur in the cities, suburbs or rural areas. While we do not have to pay for these services, when they are damaged or overused, the costs of repairing them (provided repair is possible) are often vast.

A common example is the discharge of human wastes, industrial and domestic pollutants or agricultural chemicals into rivers and oceans. Normally these wastes are treated by the aquatic or marine biodiversity—a host of species of bacteria, fungi,

Costing Ecosystem Services

Every year about $US20 billion are spent on pesticides worldwide. Studies to discover whether or not the expense is worthwhile have resulted in estimates that crop losses to pests would increase by about 10 per cent if no pesticides were used. Of course, the increases in losses would differ according to the crop. In some cases they would be zero but in others they would approach 100 per cent.

Growers all over the world have recently turned away from pesticides and towards biodiversity to save money. In some cases the direct savings have been huge. In many parts of the world, pesticide use actually increases *the pest numbers, causing even worse problems. This is because most pesticides kill the natural enemies of the pest as well as the pest itself. The pests, which generally recover faster than their natural enemies, then multiply more rapidly than usual as they are no longer subject to natural control mechanisms.*

In some regions of Indonesia, increasing levels of pesticide application did not prevent increasing losses in the rice crop and it also generated increasing human health problems. An examination of the situation resulted in the banning of 57 different pesticides. Predictions that this would cause economic ruin were proved wrong because the natural enemies (the three P's) of the major rice pest, which had previously been killed by the pesticides, then returned. Losses to pests while pesticides were in full use in the period 1986–88 were estimated at $US1.5 billion. As these losses were recovered in a few years, the value of the biodiversity that made this possible was at least that amount.

This value was for the income from the rice crop. However, other major costs of pesticide use may be eliminated by natural biological control. For example, in the United States it has been estimated that the costs of hospitalisation, outpatient care and lost work time for workers poisoned by pesticides amounts to $US25 million each year and the losses in crop pollination by the destruction of honeybee colonies by pesticides are as great as $US300 million annually. There are also costs due to groundwater contamination, fish kills and wildlife poisoning but these are often more difficult to estimate.

invertebrates and plants that break some of them down, detoxify others and recycle still more, returning them to useful ecosystem processes. However, when overloaded, large parts of this biodiversity are killed and the pollutants swill around largely unchanged. Water supplies, human health and fisheries are among the casualties. Clean-up costs may run into millions of dollars. These costs, which are rarely in any formal accounts before things go wrong, demonstrate the service aspect of ecosystem processes.

Farmers, graziers, foresters and gardeners rely on ecosystems to provide fertile soil. While most growers add fertiliser to their soils, much of the structure of the soil and fertility is actually generated and maintained by the organisms that live in it. The figures for the numbers of soil organisms are often staggering: a cubic metre of some soils contains 40,000 worms of various kinds and a similar number of insects and their relatives. Numbers of bacteria and fungi run into millions. Soil organisms process nitrogen into nitrates, recycle phosphorus and other nutrients, and break down organic matter such as dead animals and plants to make these nutrients available to the crops we grow or the animals we graze. This is the 'capital' inherited from the activities of many organisms over very long periods of time. The addition of artificial fertiliser is a bit like making a deposit into the account to maintain the capital.

As well as nutrients, crops also require protection from pests. Pesticide use has received much publicity because its financial impact on agriculture, forestry and public health is highly visible. Pesticide use has increased steadily in the last 50 years but its effectiveness often diminishes as, through natural selection, the pests evolve resistance. The suggestion that pesticide use should be decreased or stopped altogether might sound revolutionary but in many cases this strategy has had excellent results. Given a chance, the natural enemies of the pest species, which include predators, parasites and pathogens—the three P's—have kept pests under control. In the absence of pesticides, the three P's emerge from the natural biodiversity of the area and, by simply doing what they do best—attacking other organisms—they control pests.

Breakdowns in ecosystem services often cause serious problems. They are constantly in the news: a toxic bloom of blue–green algae in a river system, the demise of a coral reef, mass deaths of birds of prey or dieback of forest trees. The collapse of an ecosystem service is not intended as an experiment but that is exactly what it becomes. This is because each of these breakdowns tests the resilience of ecosystem services and, when some form of replacement or repair is necessary, we quickly discover what the costs will be. These costs take many forms: an environmental clean-up, the loss of livelihoods with compensation and retraining, the addition of environmental technology to industrial plants and the deterioration of public health. Given this list, it is as well to remember the benefits of biodiversity: clean air and water, a reliable food supply and a host of biological resources. This makes an interesting balance sheet but it is still only a small part of conventional accounting. Some countries are now beginning to include such environmental 'capital' in their national accounts.

Future Biological Resources

It is essential to manage our biological wealth soundly so it continues providing resources for future generations. In addition to its value as conventional bioresources and ecosystem services, biodiversity offers many opportunities for innovative technologies. Moreover, it has the enormous advantage of being renewable. By carefully developing it as a resource it can provide sustainable productivity.

A familiar example of the sustainable use of biodiversity is the search for biological control agents (see box on page 97). Sometimes biological pest control involves the use of natural predators, parasites or diseases. This requires exploring the native habitat of the pest for a natural enemy that best fits the agricultural system where its control is necessary.

Biological control illustrates the kind of reasoning

Clean-up costs of breakdowns in ecosystem services can be enormous. Blooms of toxic algae in our waterways are sometimes the result of industrial or agricultural run off that kills large parts of aquatic biodiversity. Restoring the biodiversity is necessary to maintain water quality.

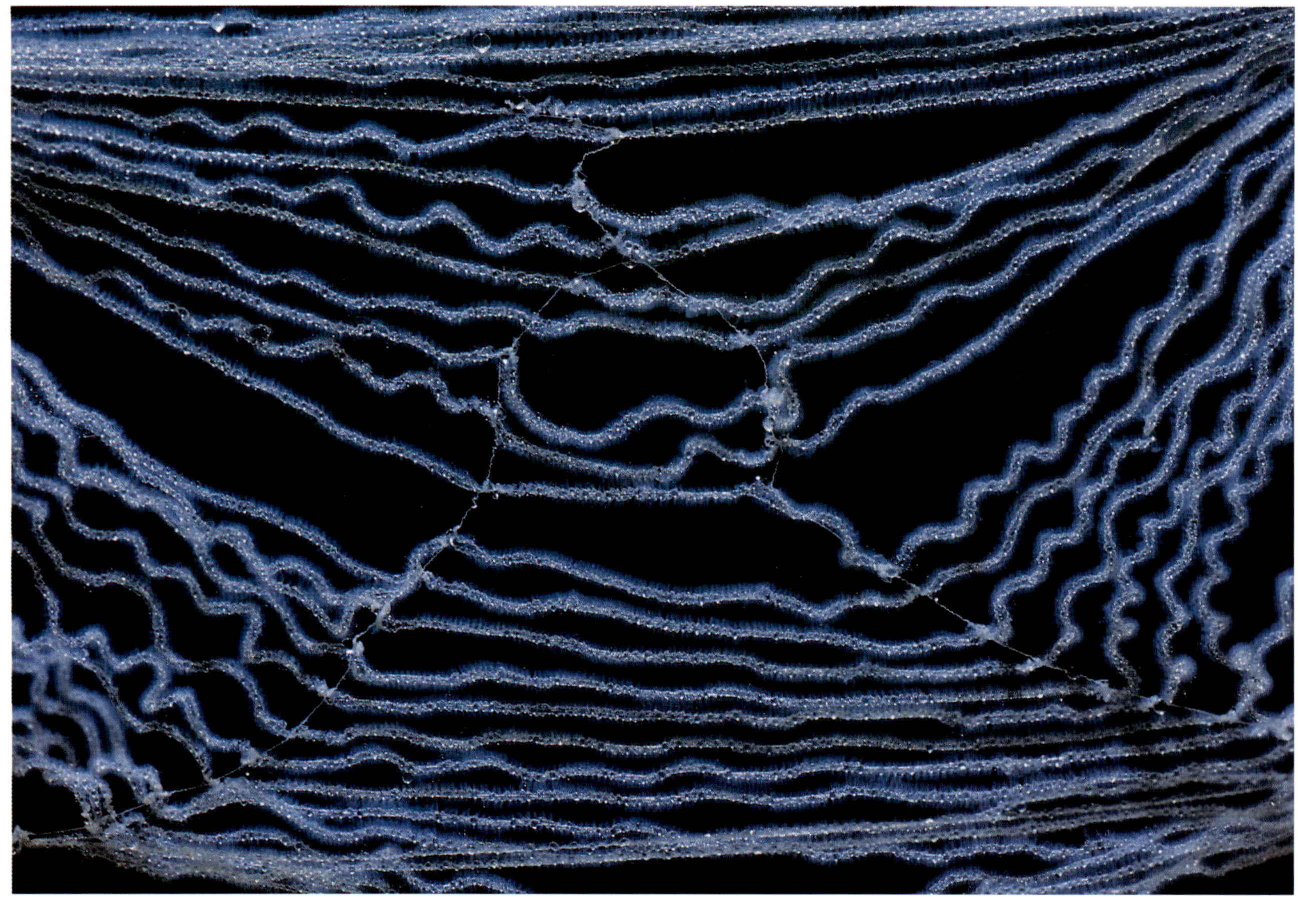

Weight-for-weight, spider silk is much stronger than steel and can be used to make very light but particularly effective bullet-proof vests. Several major companies are exploring other industrial applications for this amazing material.

that will revolutionise the search for many other kinds of biological resources: we can find wild species that already do or make the things we need. In the case of biological control, this involves finding a predator that is already adapted to killing the pest for food. The key word here is adaptation, or those aspects of the behaviour, physiology or structure of a species that enable it to survive.

We can learn a great deal from biodiversity. Most species on Earth, including humans, have many problems in common. One example is disease. In this case, the search focuses on adaptations other species have evolved to combat disease. It is then important to examine whether or not their adaptations can become our solutions. For example, we have found adaptations for fighting disease in the form of antibiotics in species as diverse as fungi and ants. Such adaptations, which have been refined over many millions of years, offer not only original and elegant solutions to human problems but many may be of considerable economic benefit. There are many examples:

• Biodiversity has been used to reduce the impact of major oil spills. Bacteria use many different kinds of chemicals as food—some of them use oil. Investigation of the microbes in oil has revealed several species that have since been developed into a 'cocktail' of oil-consuming bacteria, which is applied to affected areas. This technology is at the leading edge of a great biodiversity-based industry for the biological control of pollution.

• The armed forces need stronger, lighter fibres to stop fast-moving projectiles such as bullets and aircraft. Many spiders are adapted to capture large,

It is not necessary to spend vast amounts of money on sophisticated equipment to monitor water pollution when filter-feeders, such as these freshwater mussels, do the job naturally.

mobile prey in their webs: weight for weight, spider silk is many times stronger than steel. Spider silk, or manufactured mimics, may make superior bulletproof vests and nets able to gently stop runaway planes on aircraft carriers and commercial airstrips.

• Structural engineers searching for ways to make concrete more flexible turned to biodiversity for inspiration. It came from deep-sea snails whose shells must withstand the enormous water pressure while being flexible enough to allow the animal inside to move and grow. The micro-architecture of the snail shell has been used to devise new materials of great strength and flexibility.

• Pollution is so widespread that manufacturing and placing instruments to monitor it is often prohibitively expensive. However, many kinds of organisms sample the environment repeatedly every day of their lives. In aquatic environments different kinds of animals filter the water for food particles. In doing so they accumulate pollutants in their tissues. Underground, various kinds of worms feed on soil, extracting the nutritious particles while passively accumulating pollutants like heavy metals. Populations of these species are the equivalent of an army of monitoring instruments. By collecting and analysing these animals, the levels of pollutants are revealed over wide areas, often pinpointing the sources. Australia is a world leader in bio-monitoring technology.

Other examples of proven or potential resources include:

• New adhesives from barnacles adapted to adhere to wave-battered rocks.

• Bird-repellents for orchards derived from bugs adapted to keep potential avian predators at bay.

• The extraction of valuable minerals from low-grade ores by bacteria adapted to accumulate the mineral salts as part of their normal metabolism.

• Advances in understanding brain disease by examining the simple nerve networks of worms.

While some people argue that conservation should not be discussed in dollar terms, clearly many unexpected kinds of organisms have great potential for human welfare as well as commercial value. Humble organisms, such as bacteria, worms, barnacles, ants and spiders, are potential or proven valuable resources.

Suggested Reading

Beattie, A.J., 1994. Conservation, evolutionary biology and the discovery of future biological resources. In: Moritz, C. and Doley, D. (eds). **Conservation biology in Australia and Oceania.** Surrey Beatty & Sons: Chipping Norton.

Beattie, A.J., 1991. Biodiversity and bioresources–the forgotten connection. **Search** 22: 59–61.

Collins, D.J., Culvenor, C.C., Lamberston, J.A., Loder, J.W. and Price, J.R., 1990. **Plants for medicine: a chemical and pharmacological survey of plants in the Australian region.** CSIRO: Melbourne.

Ehrlich, P.R. and Ehrlich, A., 1991. **Healing the planet.** Addison-Wesley Publishing Company: New York.

Heiser, C.B., 1973. **Seed to civilization: the story of man's food.** W.H. Freeman: New York.

Lazarides, M. and Hince, B., 1993. **CSIRO handbook of economic plants of Australia.** CSIRO: Canberra.

Low, T., 1988. **Wild food plants of Australia.** Angus & Robertson: Sydney.

McNeely, J. A., 1988. **Economics and Biological diversity.** International Union for Conservation of Nature and Natural Resources: Gland, Switzerland.

Sandlund, O.T., Hindar, K. and Brown, A.H.D., 1992. **Conservation of biodiversity for sustainable development.** Scandinavian University Press: Oslo (distributed by Oxford University Press: Oxford).

Chapter 4

The Animals and Plants

When we think of living things, we automatically imagine the most obvious: the larger plants and animals. While both of these groups are relatively well researched, new discoveries in other groups are being made constantly and the tally of species continues to grow.

Animal Diversity

A frequently held misconception is that all animals have fur and suckle their young. However, these belong to only one small group of animals known as mammals. Animals come in a great variety of shapes and sizes, for example: sponges, jellyfish, worms, snails, insects and starfish. The animal kingdom is traditionally divided into the vertebrates and the invertebrates. The vertebrates are more familiar and include fish, amphibians, reptiles, birds and mammals, all of which have a vertebral column or backbone. By contrast, the invertebrates comprise over 30 major groups, none of which

LEFT *Some invertebrate animals. A group of sea slugs (pink) feeding on soft coral (orange). The egg masses of a sea snail (white) are prominent, centre top. The red organisms are algae.*
ABOVE *Many land invertebrates, such as these distinctive slugs, are confined to moist, dark places because their bodies lose too much water if exposed to dry air.*

Variations on the Worm Theme

A number of very diverse invertebrate animal groups are referred to collectively as worms. They share the superficial characteristics of an elongate, cylindrical and often soft, slimy body devoid of limbs—features that many people find repulsive. However, they often come in vivid colours and under the microscope display exquisite detail in body design. A worm-like shape is suited for swimming, crawling and, especially, burrowing.

The worm form as adopted by two marine creatures: a beach worm (above), a giant amongst the segmented worms (Annelida); and a ribbon worm (Nemertina), with its everted multi-branched proboscis used to trap prey (below).

Flatworms (Platyhelminthes) found in freshwater streams and in the oceans are often brilliantly coloured. Land flatworms are sometimes seen on garden paths on warm, damp evenings. Flatworms also include the parasitic flukes and tapeworms. Other parasitic groups of worms include horsehair worms (Nematomorpha), parasites of arthropods as juveniles but free-living in fresh water as adults; thorny-headed worms (Acanthocephala), parasites in both the juvenile stage (in arthropods) and as adults (in vertebrates, usually fish), and tongue worms (Pentastoma), parasites of the lungs and nostrils of mammals and reptiles. Many species of roundworms (Nematoda) parasitise animals and plants. Others are free-living in water and on land, from tropical to polar regions, and some even occur in such inhospitable places as hot springs, vinegar and the traps of pitcher plants. Tiny nematodes are also unbelievably abundant—there are possibly more individuals of nematodes than any other animal group.

The segmented worms (Annelida) include the earthworms, leeches and bristle worms. Some earthworms are aquatic but we are more familiar with those that maintain soils. Leeches occur in aquatic and land environments and while we are more likely to meet the blood-sucking type, the majority are carnivorous, feeding on small invertebrates. Bristle worms are less well known but far more diverse. They are marine, occurring from beaches to ocean trenches. Some live permanently in tubes, others actively seek their food.

This by no means completes the inventory of worms. Others include ribbon worms (Nemertina), peanut worms (Sipunculida), spoon worms (Echiura) and acorn worms (Hemichordata) as well as two groups that have no common names, the Pogonophora and the Priapulida. The former means 'beard-bearer' and the latter, 'little penis'.

A worm-like shape has even been adopted by several groups of vertebrates such as eels and snakes.

have backbones. At least 95 per cent of all animal species are invertebrates, a conservative estimate as many new species continue to be found. While the tally of vertebrates is largely complete at about 40,000 species, the number of invertebrate species is steadily climbing from the current estimate of five million.

Invertebrates are generally smaller than vertebrates, although the size range is enormous: from microscopic forms only a fraction of a millimetre in size to giant squid that grow up to 16 metres. While some invertebrates are permanently attached to a rocky shore or substrate (such as sponges, barnacles and oysters), most move about, actively seeking food and avoiding enemies. These have a distinct head end with sensory structures, like eyes and feelers, and a mouth. Most—but by no means all—animals also have a distinct back end where wastes are ejected. While animals come in a huge variety of body shapes, the worm form is very common.

Being mostly small and numerous, invertebrates form an integral part of the food web by becoming food for larger animals. Many feed by eating plant tissues or by actively hunting other animals. Others draw water through their bodies to filter out micro-organisms. This is why it is dangerous to eat oysters and other shellfish from polluted waters: harmful micro-organisms accumulate inside their bodies. A wide variety of invertebrates are scavengers, that feed on dead plants and animals or faeces providing a vital role in recycling organic matter. But even more common than all these feeding strategies combined is the parasitic way of life. Apart from micro-organisms, the invertebrates constitute the bulk of parasitic species. They include flukes and tapeworms (Platyhelminthes), roundworms (Nematoda), and many crustaceans, insects, and mites (Arthropoda). It has been conservatively estimated that 60 per cent of all insects are parasitic, meaning that parasites alone represent close to half the animals on Earth. Parasites are of immense importance in regulating animal and plant numbers.

Animals that Live in Water

Animal diversity is greatest in the oceans where the first simple, soft-bodied animals arose over 600 million years ago. Fossils of these forms have been found at Ediacara, in the Flinders Ranges, South Australia, where there was once an inland sea. Present day marine biodiversity is vast. It ranges from animals specialised to life on the surface to bizarre creatures that live at great depths where no sunlight penetrates. Recently, new kinds of animals have been discovered at hydrothermal vents—places where undersea volcanoes are active (see box on page 58). Aquatic animals include sponges (Porifera), sea anemones, jellyfish and corals

The greatest diversity of animal life occurs in the oceans. Many groups of animals, such as octopuses and their relatives—squids and cuttlefishes—are exclusively marine.

Peripatus are unusual creatures that represent a living link between the segmented worms (Annelida) and centipedes, millipedes and insects (Arthropoda). They also have a long evolutionary history, with fossil forms over 300 million years old that are very similar to present-day species.

(Cnidaria), sea stars and sea urchins (Echinodermata). Less well-known groups include the lamp shells (Brachiopoda), which superficially resemble clams; moss animals (Ectoprocta), which have a plant-like growth, and arrow worms (Chaetognatha)—tiny torpedoes with voracious appetites that dart around just below the surface of the ocean. Sea squirts (Urochordata) are animals encased in a leathery skin and permanently attached to the rocky shore. Surprisingly they are closely related to the vertebrates. This can be seen during their tadpole-like larval stage, which possesses a supporting rod called a notochord down the back, the forerunner of the backbone.

Animals that Live on Land

Descendants of the transitional forms that started colonising the land are still living today as 'missing links'. Around 400 million years ago, several groups of fishes living in poorly oxygenated stagnant pools developed ways of breathing air at the surface. Descendants of these fish are the lungfish that occur in Africa, Australia and South America. It is from such fish-like ancestors that the Amphibia, the first of the land vertebrates, arose. As a remnant of their past diversity the Amphibia occur today as frogs and toads, salamanders and a small but curious group of burrowing legless animals called caecillians.

Over time, reptiles evolved eggs with shells that prevented the embryos from drying out and so became capable of living away from water. Today, reptiles are represented by lizards and snakes, alligators and crocodiles, and tortoises and turtles. Following the extinction of much of the diversity of the reptiles, including the well-known demise of the dinosaurs about 65 million years ago, birds and mammals became the dominant vertebrates on land. Most have bodies covered by an insulating layer of feathers or fur, they can maintain a high and constant body temperature, making it possible to remain active as the temperature fluctuates.

At the same time the first land vertebrates appeared, terrestrial peripatus or velvet worms (Onychophora) evolved from an ancient marine form. These soft, caterpillar-like animals share characteristics with both worms and arthropods. They are still living today in Australasia, Africa, South–East Asia and South America in damp places such as rotting logs and leaf litter.

On land, many invertebrate groups are confined to moist, dark places; others survive in the driest deserts. The majority, however, occur in forested regions, particularly in the tropical rainforests. In the soil, leaf litter and rotting logs on the forest floor live flatworms (Platyhelminthes), earthworms and leeches (Annelida), and snails and slugs (Mollusca), all requiring moisture because their bodies desiccate if exposed to dry air.

The two groups of animals that can best withstand dry conditions are the insects and their relatives (arthropods) and the terrestrial vertebrates.

A lesser-known group of arthropods is the Tardigrada. These animals are common but microscopic. The number of species in Australia is unknown.

The Arthropods

Arthropods are the most abundant and diverse group of animals on Earth. Approximately one million have been described, about three-quarters of all known animal species. They range in size from microscopic forms to the giant Japanese Spider Crab with a leg span of four metres. They have in common the distinctive feature of a skeleton that is worn like a suit of armour on the outside of the body. The legs and other appendages have joints to allow movement—the word arthropod derives from Greek, meaning 'jointed legs'.

The appendages are arranged in pairs down the length of the body and while some are used for walking or swimming, others are modified as antennae (feelers), jaws, syringes for piercing the surface of plants or other animals, as hands for manipulating food or as devices for copulation.

Arthropods come in several basic models. The biggest group includes the centipedes and millipedes, which have many legs, and the insects, which have just three pairs confined to the thorax, which may also sport wings.

Crustaceans are largely confined to the sea and fresh water, but some (slaters and sandhoppers) occur in damp places on land. While the larger crustaceans (prawns, lobsters and crabs) are most familiar because they are tasty, the vast majority are much smaller. In the Antarctic, small shrimp-like crustaceans known as krill occur in vast numbers and form the diet of the Blue Whale, the largest animal that has ever lived, which can eat up to four tonnes of krill per day.

The last group includes spiders, scorpions, ticks and mites: the arachnids. Their most distinctive feature is that they have four pairs of walking legs. They do not possess antennae and the front appendages are modified for feeding, such as poison fangs in spiders, pincers in scorpions and piercing and sucking beaks in mites and ticks.

Only a fraction of arthropod biodiversity has been documented. At one time it was thought that the insects had the most species but it may be that the mites have as many or more. Mites utilise an astonishing variety of habitats including the bodies of most plants and animals.

Cycads are among the most ancient land plants, differing little from their ancestors of the Triassic Period, some 200 million years ago. They occur across northern Australia and along the east coast.

Consequently these two groups are dominant on land. Both have a substantial skeleton to support them, the arthropods wearing it on the outside as an exoskeleton. Both groups have bodies covered by a thin layer of wax to reduce water loss from the skin. You might notice how your skin becomes wrinkled after a long bath: the soapy water dissolves this waxy layer, removing your natural waterproofing. Respiratory systems are internal, with air-filled tubes in the arthropods and lungs in the vertebrates. But how could these groups avoid desiccation when it came to reproduction? Sperm and eggs are delicate cells that dry out easily. In the aquatic environment, they can be safely released into the water where fertilisation takes place. However, land arthropods and vertebrates avoid the hazard by introducing sperm directly into the body of the female. The developing embryos are also protected from desiccation either by the provision of a waterproof shell, as in insects, reptiles, monotremes and birds, or by the retention of the embryo inside the body of the mother until birth.

Plant Diversity

Green plants range from single-celled algae to trees as high as twenty-storey buildings. They occupy all kinds of habitats including the coldest, hottest and driest on Earth and are found wherever there is sunlight: in fresh water, the oceans and on land. The different sizes, shapes, structures and colours of plants present a dazzling array but despite the dramatic variation in appearances, they are all able to use sunlight to produce energy, through the process called photosynthesis (see box on page 39). Plants are divided into five basic groups. The most familiar of these is the flowering plants, which currently comprises more species than all the other groups put together. They grow in a very wide range of habitats: from desert to rainforest, polar regions to tropics and dry land to estuaries. They grow rooted in soil or water or in places where they can get hold, such as on the branch of another plant. Known as angiosperms, flowering plants include woody shrubs, trees, vines and herbaceous plants.

The gymnosperms include cycads, which have thick woody trunks with crowns of large divided leaves, and conifers, such as the Norfolk Island Pine.

Ferns and related plants grow in soil, rock crevices or perched in the branches of trees. Grouped under the name pteridophytes, they range in size from a few centimetres to several metres tall. Some tiny ferns float in rivers and billabongs, while tree ferns have tall trunks and crowns of spreading fronds.

Mosses and their relatives are known as bryophytes. These are usually small plants, ranging from a few millimetres to about 60 centimetres. Most live on soil or rock surfaces; a few grow in freshwater creeks, ponds or lakes and many grow on the surfaces of other plants. In very moist habitats like rainforests, bryophytes cover bark on tree trunks, festoon branches and sprout from leaves. They can even grow on the backs of beetles.

Algae form an immense and diverse group of plants that contains over 22,000 recognised species. They range in size from a single cell up to gigantic fronds of sea kelp 100 metres long. Many algae are aquatic, living in ponds, lakes, creeks and rivers. Dense seaweed forests as complicated as the forests on land grow in the oceans along the continental shelf. Other algae live on land in moist places in soil, rocks, tree trunks, fence posts, the crevices of desert rocks and even in the polar regions.

All plants contain a range of coloured pigments able to trap light energy at particular wavelengths for photosynthesis. While the colours vary little between land plants, algae exhibit an extraordinary range. Some pigments enable certain algae to live at ocean depths where sunlight is very weak and only feeble light at the blue end of the light spectrum penetrates the water.

Plant Reproduction

Many plant groups share a fascinating reproductive cycle. This cycle always begins with a familiar step: the fertilisation of an egg by a sperm.

In ferns, which are among the most ancient land plants, fertilisation produces the familiar leafy plant. When mature this plant produces spores in special structures on the leaves. Germination of the spores produces not another fern plant but a small scrap of green tissue, about the size of a five-cent coin. This tiny plant is independent, feeding itself by photosynthesis. (In some relatives of ferns, independence is maintained by teaming up with a fungus that provides nutrients.)

This tiny structure then produces sperm and eggs. After fertilisation, the eggs grow into leafy plants, starting the cycle all over again. So although it may be difficult to believe, the ferns in your garden or pot plant lead a double existence: the familiar spore-producing plants and the little-known plantlets that bear eggs and sperm.

Although more advanced plants have evolved

Some plants, like ferns, produce spores from cases on the underside of their fronds. From the spores, a tiny plantlet grows that produces eggs and sperm from which adult plants are produced.

much larger and more complex bodies with roots, trunks, branches, leaves and flowers, traces of this reproductive cycle lie hidden in some familiar structures. In the flowering plants, the sperm and eggs are borne on separate plantlets that are both contained within the flower. Concealed inside the ovary, the eggs are surrounded by a few cells that are the remains of what was, in more ancient plants, the independent plantlet. Evolution has increased the protection of the eggs by retaining them within the large, spore-bearing plant, eliminating the exposed and vulnerable scrap of tissue on the ground.

Reproductive organs in flowering plants are housed within the flower such as this water lily. The eggs are enclosed inside the ovaries and the sperm are encased within the tiny pollen grains.

The remnants of the male plantlet are so small that they fit, together with the sperm they produce, inside a pollen grain. This miniaturisation enables the sperm, each carefully protected, to travel from one flower to another on wind currents, or to be transported on the bodies of pollinating animals, such as possums, flying foxes, birds or, especially, insects like bees. Pollen grains are produced in huge quantities and many of those borne on the wind end up in strange places, including noses, as all those who suffer from hay fever know.

Following fertilisation, the eggs develop into seeds within the ovary. This then becomes a fruit, protecting the seeds during their development to maturity. Seeds may be dispersed by gravity, wind, or by being eaten and later defecated by an animal attracted by the fruit. Seed-dispersing animals include many kinds of birds and mammals, and also, curiously, various species of fish and some ants.

Flying foxes are important pollinators and seed dispersal agents of native plant species.

Suggested Reading

Bell, P.R., 1991.
Green plants: their origin and diversity.
Cambridge University Press: Cambridge.

Bennett, I., 1992.
Australian seashores: a guide to the formation, animal and plant life.
Collins Eyewitness Handbooks, Harper Collins: Sydney.

Brusca, R.C. and Brusca, G.J., 1990.
The invertebrates.
Sinauer Associates, Inc. Publishers: Sunderland, Massachusetts.

CSIRO Division of Entomology, 1991.
The insects of Australia (Volumes 1 and 2).
Melbourne University Press: Melbourne.

Encyclopedia of Australian Animals, 1992.
In four volumes: **Mammals**, by R. Strahan; **Birds** by T.R. Lindsey; **Reptiles** by H. Ehmann and **Frogs** by M.J. Tyler.
The Australian Museum/Angus & Robertson: Sydney.

Ingrouille, M., 1992.
The diversity and evolution of plants.
Chapman & Hall: London.

Morley, B.D. and Toelken, H.R., 1983.
Flowering plants in Australia.
Rigby Publishers: Adelaide.

New, T.R., 1992.
Introductory entomology for Australian students.
University of New South Wales Press: Sydney.

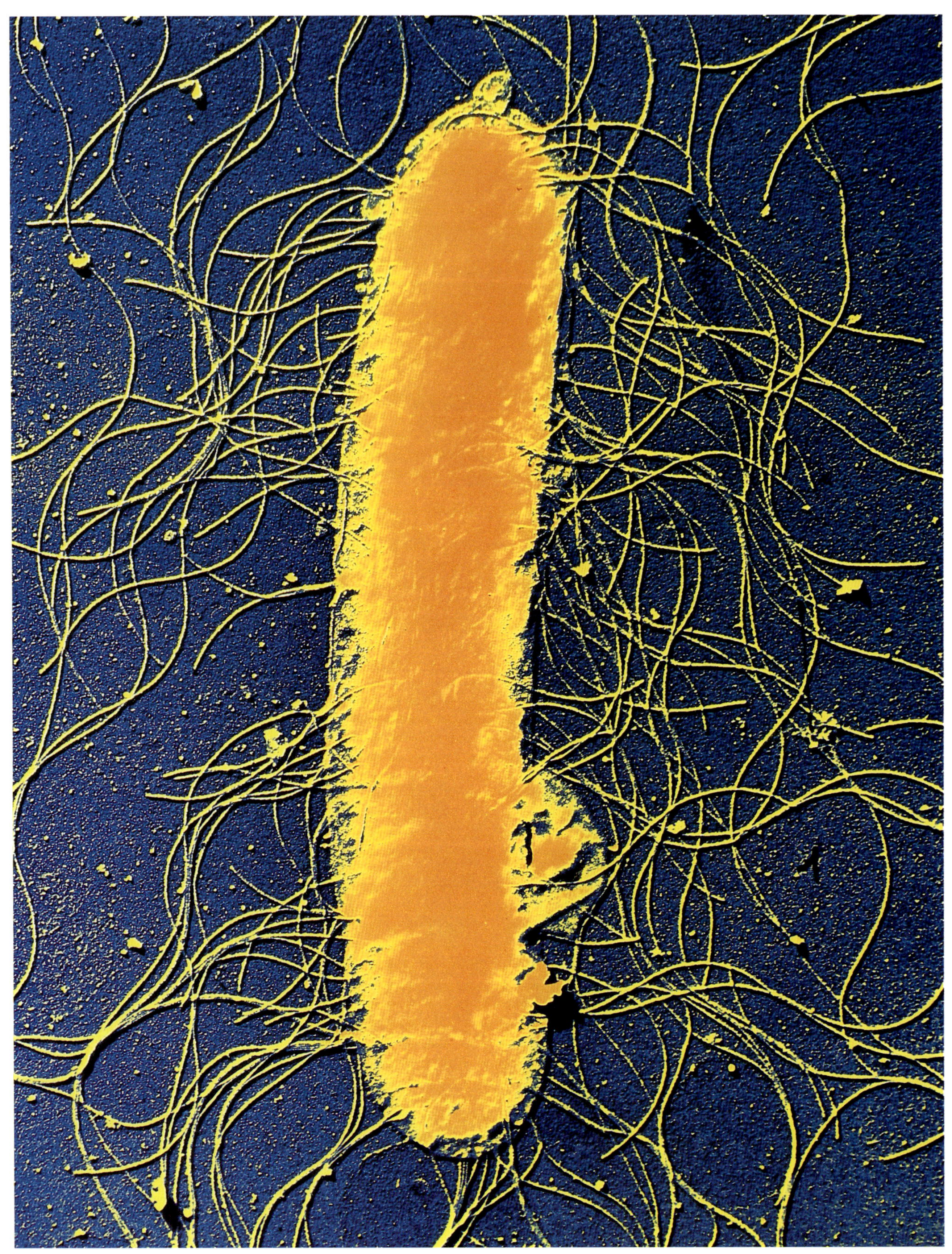

The Microscopic Life Forms

Although micro-organisms are too small to see with the naked eye, they probably represent the greatest diversity of all living things. The tiny size of micro-organisms meant our awareness of their existence had to wait until the development of the microscope in the 17th century. But it was not until the latter part of the 19th century that the specialised techniques required to study microbiology were developed. Because microbiology is so young, our knowledge of micro-organisms lags way behind that of other groups.

What is a Micro-organism?

Micro-organisms are defined by their size. They are simply life forms impossible to see without the aid of a microscope. These tiny living things include algae, fungi, protozoa, Cyanobacteria (blue–green algae), bacteria and viruses.

Like the vast majority of bacteria associated with the human body, Proteus mirabilis *lives as a harmless passenger in the human intestine. Other gut-associated bacteria produce vitamins essential to human health.*

Although size is a convenient way to define micro-organisms as a group, it conceals very large differences in their structure. The greatest structural difference between micro-organisms is in their cells. In the eukaryotic cell type, the tiny organs within the cell—more correctly called organelles, including the nucleus where the DNA is stored—are surrounded by special membranes. This is the cell type found in fungi, protozoa and algae as well as plants and animals. By contrast, the bacterial and cyanobacterial cells are simpler, with no membrane-bound organelles. These are called prokaryotic cells. These differences may seem trivial but they are actually greater structural differences than those between animals and plants.

Where are Micro-organisms Found?

The numbers of micro-organisms found anywhere in the environment can be staggering. A gram of fertile soil—the size of two aspirin tablets—may

The Biodiversity of Hydrothermal Vents

Perhaps no other area of this planet has been less explored than the deep ocean floor. This region presents a seascape of vast plains, mountains, ridges and trenches several kilometres below the ocean surface. Until recently, the physical features of these areas were believed to be at the limits of life support and therefore relatively barren. However, exploration has revealed a wealth of strange creatures capable of surviving great pressures, extremes of temperature and pitch darkness.

A spectacular example of this diversity has been discovered in deep mid-ocean ridges such as the Galapagos Rift in the Pacific off the coast of South America. Here volcanic activity is creating a new ocean floor. Like volcanoes on land, the material spewed from the vents, known as 'black smokers', contains an abundance of rotten egg gas (hydrogen sulphide). Bacteria, using the energy derived from the oxidation of hydrogen sulphide, convert carbon dioxide into organic compounds to build their bodies. Thus, in total darkness, these bacteria support a complete ecosystem in the same way plants do with light energy.

These bacteria form the base of the food chain for a wealth of animal life. Hydrothermal vents have revealed 16 families of invertebrates that were unknown just five years ago. The animals that live there include clams, crabs, segmented worms and fish. The most unusual is a group of tube-living worms measuring over a metre in length and recently placed in a completely new phylum, the Vestimentifera (see 'What are Phyla?' box on page 76). These worms have no intestine but cultivate a garden of bacteria in special regions of their bodies for their nourishment. The bacteria are supplied with hydrogen sulphide and oxygen from the blood of the worms.

Even more recently a huge field of vents has been discovered lying along the mid-Atlantic Ridge, west of the Azores, at depths of around 1,600 metres. These vents harbour a new set of species, entirely different to their Pacific counterparts.

contain 100 million bacteria as well as a myriad of other micro-organisms. Each bacterium measures only about a thousandth of a millimetre. Despite their tiny size, micro-organisms are thought to constitute between five and 25 times the total mass of all other life, both in water and on land.

Even the human body is not the single organism it appears. Ninety per cent of the total number of cells in your body are not human but are, in fact, bacteria. These bacteria are present in vast numbers on skin, hair and teeth, and in your throat, nose and intestine. Most of them are harmless but some can cause disease. Many are beneficial, making essential vitamins or controlling the growth of other, disease-causing micro-organisms. We are also hosts to a numerous other micro-organisms, including fungi and protozoa.

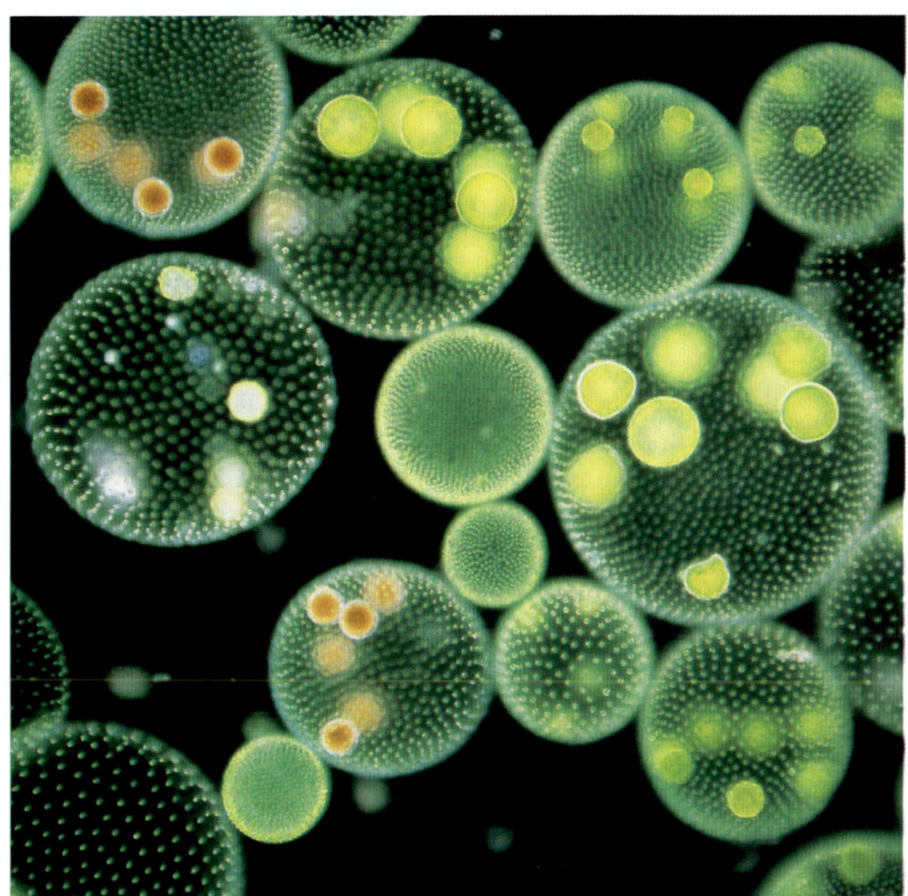

Cells within a Volvox *colony, a green alga, act in a cooperative fashion so that the entire colony behaves like a single organism. Tiny hair-like structures called flagellae, beat in a coordinated manner to move it through the water.*

Micro-organisms multiply rapidly. Given ideal conditions most bacteria will double in number every 20 to 60 minutes. Some can even double their population every 10 minutes. Within 48 hours a single bacterium with a typical doubling time of 25 minutes would increase in biomass to an amount roughly equivalent to 4,000 times the mass of the Earth! Fortunately factors such as lack of nutrients prevent unlimited microbial growth.

Micro-organisms define the limits of where life can exist on Earth. They have been isolated drifting on air currents at heights of up to 32 kilometres and at ocean depths as great as 11 kilometres. Most life is confined to temperatures between –5°C and 50°C but heat-loving or thermophilic bacteria flourish at temperatures that scald or kill larger organisms—in hot springs, burning coal tips and compost heaps. Some thrive at temperatures close to the boiling point of water. In some oil wells and around deep-sea hydrothermal vents, high atmospheric pressure raises the boiling point of water to above 100°C. In such environments thermophilic bacteria thrive at temperatures of 105°C. At the opposite extreme, cold-loving or psychrophilic bacteria grow at temperatures as low as –8°C. Such organisms cause spoilage of refrigerated foods.

Some micro-organisms, like us, require oxygen. These are called aerobes and they use oxygen to break down food for energy, producing carbon dioxide and water in the process. Micro-organisms that do not require oxygen are called anaerobes.

They gain their energy using a process known as fermentation. This can yield a variety of products, the most notable being alcohol.

There are other micro-organisms that, instead of using oxygen, use other highly oxidised chemicals such as sulphate or nitrate. These bacteria are called the nitrate- and sulphate-reducing bacteria. The sulphate-reducing bacteria are responsible for the rotten egg smell (hydrogen sulphide) that comes from some stagnant ponds and marshes.

How Many Micro-organisms are There?

The exact number of micro-organism species is unknown but evidence indicates millions remain unidentified. There is no central catalogue of micro-organisms but *Bergey's manual of systematic bacteriology*, the major catalogue of bacteria, describes 40,000 species. Most microbiologists believe this represents the tip of a huge iceberg.

It is difficult to assess how many species of micro-organisms there are simply because their tiny size requires specialised techniques to study them. These techniques were not formulated until late last century and suitable methods have yet to be developed for most. With bacteria, for example, identification and classification have, until very recently, depended on our ability to grow them. If we cultivated the micro-organisms from a one-gram sample of backyard soil we might reveal a million, mostly bacteria. But if we re-examined that same gram of soil directly, using a microscope, we will find about 100 million micro-organisms. So cultivation techniques make it possible to study only about one per cent of micro-organisms.

There is yet another obstacle. Even those micro-organisms we can grow are frequently difficult to identify to species level. This is because, by definition, a species is a group of organisms that exchanges genetic information only with its own kind. Bacteria do not fit this definition because they can exchange genetic information across species—even very distantly related ones. Unfortunately, it is very difficult to observe this exchange process. So scientists have assumed that, to be in the same species, two individuals must have genetic information (DNA) that is 70 per cent identical. This is a very arbitrary figure. If applied to us it would classify humans, apes and monkeys as a single species.

Recent technological advances should clear up many of these problems. In the last 15 years developments in molecular biology have made it possible to investigate the DNA of organisms directly and to assess diversity by looking at the variety of DNA in a sample. It is now even possible to extract and decipher genetic material directly from micro-organisms in the environment. These techniques allow us to study micro-organisms that we are unable to grow in the laboratory.

Two independent molecular studies have started to confirm the tremendous diversity microbiologists suspected. Both examined microbial diversity in well-characterised microbial communities and revealed that none of the organisms found using molecular techniques were the same as those using traditional culturing techniques. This showed that the diversity in natural populations is far greater than previously recognised. In a separate study, molecular techniques were used to assess microbial diversity in beech-forest soil and marine sediments. These studies found 4,000–5,000 different bacterial species in a single gram of sediment taken from each environment, with almost no overlap.

This raises an awesome question: if 9,000 different types of micro-organisms exist in two grams of sediment taken from two separate locations, what unimaginable number of micro-organisms exist planet-wide?

The Economic Importance of Microbial Biodiversity

Micro-organisms are crucial for solving some of our most pressing environmental and economic problems. For example, Australia loses two-and-a-half million tonnes of nitrogen, a vital nutrient, from its farms annually. It is either consumed by the crops and livestock, leached from the soil by rainfall or lost as a gas. Nitrogen fertiliser replaces only around 300,000 tonnes. Nitrogen obtained from the atmosphere by bacteria provides much of the remainder required (over two million tonnes). At a cost of $1,000 per tonne for fertiliser, this amounts to a considerable saving to Australian rural industries. Bacteria are the only organisms known to obtain nitrogen from the atmosphere. As Australian soils become more acidic or more saline—changes that occur through agricultural practices—it becomes increasingly urgent to search for new species of nitrogen-fixing bacteria that can tolerate these changes.

Micro-organisms also maintain soil fertility by recycling mineral nutrients essential for plant growth. A common but specialised group of fungi called mycorrhizal fungi have a particularly intimate association with plants. They attach to the roots and enhance the uptake of water and vital nutrient minerals, especially phosphorous. This association is particularly important in many Australian soils, which are often deficient in phosphate.

Micro-organisms are also important in preventing soil erosion. They bind soil particles together either mechanically or with gummy substances they produce. Soil erosion is a massive problem in Australia. It probably affects 90 per cent of arable land and loss of soil structure is estimated to cost $100 million annually in the Murray–Darling Basin alone.

We are also heavily dependent on micro-organisms to alleviate pollution, both in specialised treatment facilities and in natural environments.

New technologies are emerging that enable us to use micro-organisms to break down many apparently intractable wastes, such as some pesticides, herbicides and industrial chemicals. Such technologies depend on searching the vast biodiversity of micro-organisms for pollutant-destroying bacteria. A 50-hectare polluted site on the shore of Sydney Harbour demonstrates the economic value of this technology. Its value was assessed at $15 million when contaminated. Following decontamination by micro-organisms, its value will rise more than ten-fold.

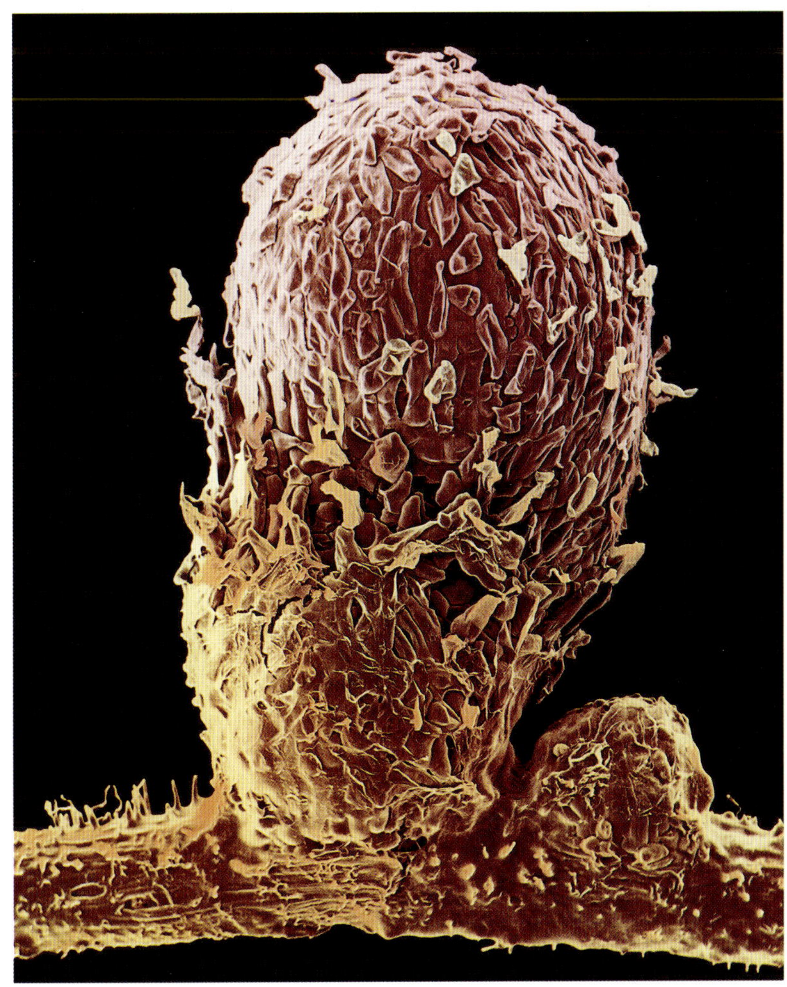

Only certain bacteria can convert nitrogen gas in the atmosphere into chemicals useful to other species, a process called nitrogen fixation. Most legumes, such as pea plants, provide these specialised root nodules to house the bacteria, which provide the plant with nitrogen.

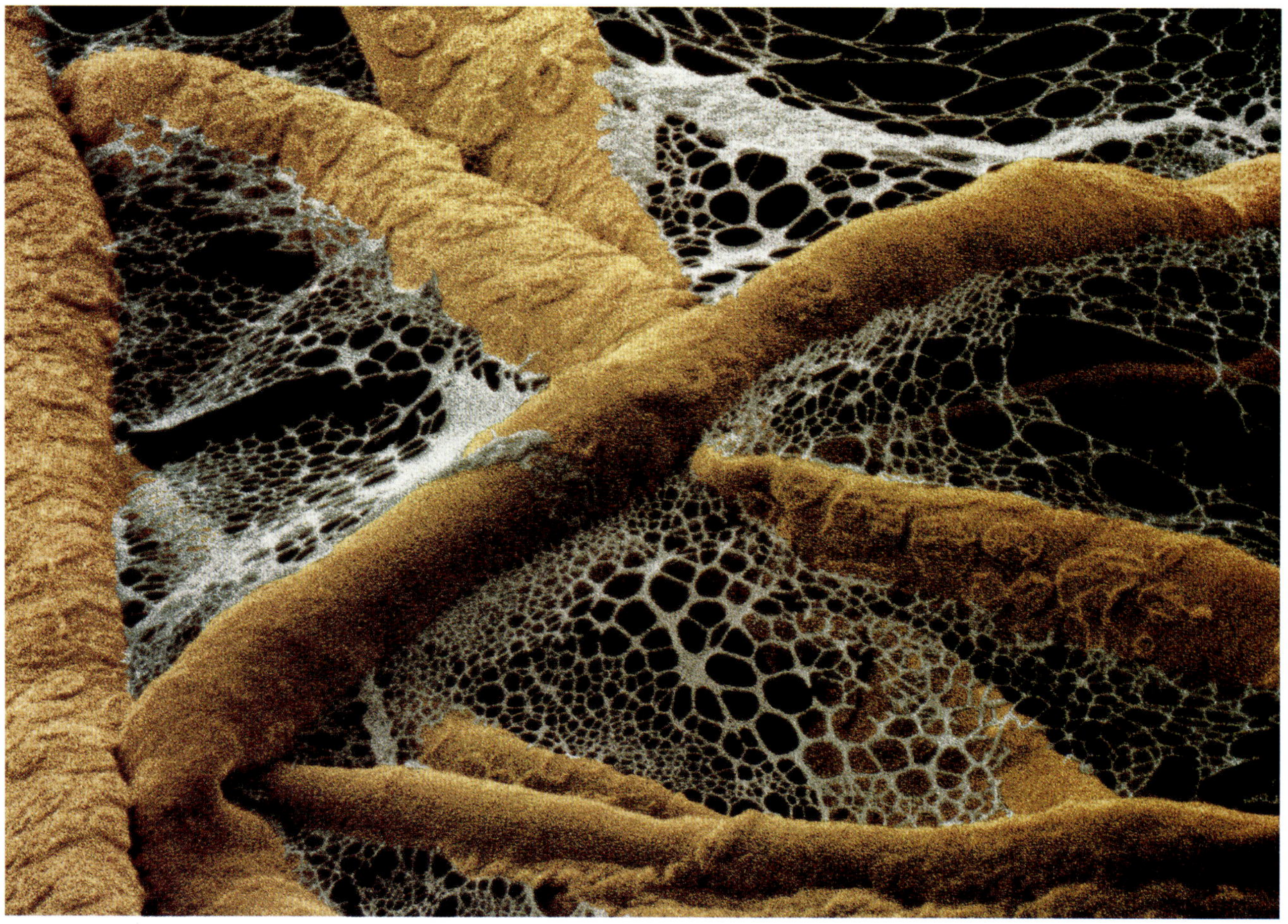

Mushrooms and toadstools are the fruiting structures of much larger organisms. The bulk of the fungus, which consists of kilometres of fine threads called hyphae, normally remains unseen as it grows throughout its substrate. This scanning electron micrograph shows fungal hyphae, tinted grey, growing in wood.

Micro-organisms form the basis of huge industries. For example, well over 3,000 biologically active compounds have been identified in one group of micro-organisms alone, the Streptomyces. These include antibiotics and anti-parasitic drugs and have been responsible for the development of a multi-million-dollar pharmaceutical industry. Yet, so far, only a tiny fraction of the Streptomyces has been explored for sources of useful compounds.

The Conservation of Microbial Biodiversity

Food sources, temperature, barometric pressure, acidity, oxygen and moisture are the most crucial factors in the habitats of micro-organisms, which respond in subtle ways to the complex chemistry and physics of their tiny worlds. Microbial diversity reflects the changes and combinations of these factors but, at present, we have no idea how. Therefore, the only feasible way to preserve microbial diversity is to preserve the diversity of their environments (see Chapter 8).

Microbial Groups

Fungal Biodiversity

The fungi have their own kingdom because they are a distinct group that is neither plant nor animal but has features of both. They resemble plants because they do not move and have rigid cell walls. Often these cell walls are made of cellulose, the major component of plant cell walls. Other fungi use chitin—the same material found in insect

Micro-organisms as Partners

Many bacteria and fungi live in intimate association with other organisms. Many of these associations are thought to be mutualistic, that is, both of the partners benefit from the association in some way.

Lichens look like plants but their tissues are actually a mixture of fungi and algae. The algae use light energy to photosynthesise and the fungi absorb mineral nutrients. It is unclear precisely how the algae benefit. Possibly the fungus helps transport nutrients and protects the alga against environmental factors, like drying out. Lichens grow very slowly—some on the Canadian tundra could be 4,500 years old. Because most lichens are resistant to temperature extremes and to drying, they can grow in hostile habitats like rock surfaces and in Antarctica. Lichens produce a range of unusual chemicals, many of which have potential applications in medicine. They are also useful in the detection of airborne pollutants, providing biological monitoring of air quality.

Leguminous plants like clover, peas, beans, peanuts and wattles have a special relationship with the nitrogen-fixing bacteria of the genus Rhizobium. *The plant provides food to the bacteria, in turn, benefiting from its own private supply of nitrogenous fertiliser. Legumes house the bacteria in special nodules, seen as lumps on the roots. An exciting area of research is to find ways of establishing nitrogen-fixing bacteria in the roots of the major crop plants of the world such as rice, wheat and corn so they do not need the application of nitrogen fertiliser.*

Some fungi also play a special role in the nutrition of plants. Known as mycorrhizal fungi, they penetrate the roots of many species of plants in such a way that they transfer nutrients from the soil into the plant tissues. The fungus benefits because the plant provides it with food. Many types of forest depend on the presence of mycorrhizal fungi on the roots of the trees. They are so crucial to many orchids that if their seedlings are not infected by the right fungus, they cannot grow.

Surprisingly, most animals that eat only plants are unable to digest the cellulose they contain. Many have overcome this problem by using micro-organisms to digest it for them—sheep and cattle cannot digest grass properly without thriving communities of microbes in their guts. Likewise, fungus-gardening ants and termites make piles of cut leaves and then infect them with a particular fungus that breaks down the cellulose. These fungal-gardeners eat the fungus, which has absorbed the nutrients in the leaves, but they always leave enough to infect more gardens.

Lichens are a living collaboration between algae and fungi.

exoskeletons. Fungi differ from plants because they lack chlorophyll so they cannot photosynthesise. Like animals, fungi depend on preformed food sources for their nutrition. They obtain this food from both living and dead organisms. When their food is living, fungi are parasites or predators; when their food is dead they are decomposers. Fungi can use a variety of food sources, including wood. This is important in terms of recycling of nutrients in forests but may be disastrous for the wood in our houses.

Mushrooms, toadstools, puffballs and bracket fungi are all reproductive structures borne by microscopic networks of fibres that make up the bulk of the organism. This network, the mycelium, grows throughout the food source before the reproductive structures appear. For example, a bracket fungus we see on the side of a tree trunk is actually the product of a mycelium that may have tunnelled its way through kilometres of tree tissue in the trunk, branches and roots. Fungi also cause disease in humans including athletes foot, thrush and ringworm.

Yeasts are fungi that grow as single cells. They may be as small as five-thousandths of a millimetre in diameter and some species are used in the manufacture of bread and alcoholic drinks.

There are approximately 80,000 known species of fungi but recent estimates are of well over a million species. They occur in all environments, including tropical rainforests, oceans and deserts. Different fungi can grow in extremes of acidity, at high sugar and salt concentrations and at low temperatures. These features make them particularly important in the spoilage of foods. We often try to inhibit fungi in food by making it more acidic (vinegar pickles), increasing the sugar (jams and preserves) or salt concentration (salted meats) or by storing at low temperatures.

Protozoan Biodiversity

Protozoa are single-celled organisms and because they do not have a rigid cell wall like plants, they have mostly been regarded as single-celled animals. Protozoa are unable to photosynthesise but some incorporate single-celled algae into their bodies. They are among the most fascinating and beautiful groups of organisms when examined under the microscope and exhibit a wide range of cell structures and forms. Amoebae found in pond water are protozoa that are just visible to the naked eye.

The protozoa form a very diverse group estimated conservatively to contain 100,000 species. Most environments have them and they play important roles in food chains by feeding on bacteria. In turn, they are eaten by a huge array of invertebrates. Some protozoa cause disease, including malaria and dysentery.

Bacterial Biodiversity

Under the microscope, most bacteria look like tiny rods, spheres or spirals. However, this simplicity of shape does not mean that bacteria lack diversity. On the contrary, the diversity of their physiological activities and the materials they use as sources of nutrition is immense. They use a vast range of materials including gases, such as hydrogen sulphide and ammonia; solids, such as sulphur and plastics, and some can even use the very antiseptics we use as disinfectants. Here we can provide only a brief insight into the diversity of bacterial nutrition.

Organisms require a source of carbon, the major building block of all life forms, plus an energy source, which is used to drive the building process. Bacteria, as a group, are very versatile and gain their carbon and energy from remarkably diverse sources. Some bacteria gain energy from organic foods in a similar manner to animals. When bacteria use dead material as a food source it is called decomposition; when using living material, such as human tissue, it is called disease.

RIGHT *Protozoans are single-celled organisms and are found in most environments. These* Vorticella *are attached by stalks to plant debris.*

Numerous materials are subject to decomposition by bacteria including wood, leaf litter and dead plants and animals. This means, of course, they are an integral part of the recycling of nutrients in all natural communities. In addition, some can use toxic chemicals such as pesticides and PCBs as sources of nutrition. We are completely dependent on microbial decomposition processes to remove toxic chemicals and to recycle essential nutrients.

Bacteria are also capable of causing disease in virtually all living organisms, including other bacteria. However, it is important to note that only a very tiny fraction of bacterial diversity consists of species that cause disease.

When plants photosynthesise they obtain energy from sunlight and carbon from carbon dioxide in the atmosphere. Many bacteria use a similar process. Also, when plants photosynthesise, oxygen is produced. Photosynthesis by Cyanobacteria (blue–green algae) also produces oxygen and it was this process that was responsible for increasing the oxygen in the Earth's atmosphere to levels that permitted the evolution of most modern life forms. Other bacteria are fundamentally different to the Cyanobacteria and plants and do not evolve oxygen during photosynthesis. Some obtain their energy from inorganic chemicals such as ammonia or hydrogen sulphide, something that no animals or plants can do. These bacteria are known as lithotrophs. Some lithotrophs gain carbon from carbon dioxide while others gain their carbon from organic sources.

Fixation of carbon dioxide may be similar to that found in plants or it can be a fundamentally different biochemical process. The sulphur bacteria use compounds such as pyrite (iron disulphide), hydrogen sulphide gas and sulphur. The

This highly saline lake in Western Australia is pink due to the growth of salt-loving bacteria containing a red pigment.

activities of these bacteria can generate quite concentrated sulphuric acid that causes both corrosion and pollution problems. For example, when pyrite is exposed during the mining of coal and metal ores, the acid produced by these bacteria makes heavy metals from the mine workings water soluble. The heavy metals in turn pollute the rivers. This is not a trivial problem: the Rum Jungle Mine in the Northern Territory released 130 tonnes of copper, 100 tonnes of manganese, 40 tonnes of zinc and 13,000 tonnes of sulphate into the Finnis River system in one year alone. However, the same sulphur bacteria can be used to recover metals from mine wastes or to remove sulphur from coal.

The activities of the sulphur bacteria have received considerable attention as a result of finding highly productive ecosystems in the deepest trenches of the Pacific Ocean and off the Galapagos Islands. These ecosystems are associated with volcanic, hydrothermal vents, which spout hot water rich in hydrogen sulphide (see box on page 58). Unlike familiar ecosystems based on photosynthesis, these ecosystems depend on the oxidation of sulphur compounds by the sulphur-oxidising bacteria. The sulphur bacteria have even formed mutually beneficial relationships with clams and worms that depend entirely on the bacteria for food. The creatures that live around these vents are mostly unique and were unknown until the discovery of the vents in the mid-1970s. Other lithotrophic bacteria gain energy from compounds like hydrogen gas and methane.

New techniques in molecular biology enable the relationships between different organisms to be established directly from their genetic codes. On the basis of these techniques microbiologists now argue that there are two fundamentally different kinds of bacteria: the Archaebacteria and the Eubacteria. It has been suggested that these should be two kingdoms. The criteria for distinguishing between them are extremely technical and do not concern us here. However, molecular criteria also suggest that animals, plants and some micro-organisms such as fungi, protozoa and most algae could be lumped together in a third kingdom, the Eukaryota. The molecular evidence suggests that the Archaebacteria and Eubacteria are no more closely related than either is to the eukaryotes and that all other organisms, no matter how different they appear to us, belong in the third kingdom. As discussed in Chapter 1, the matter of how many kingdoms there are is not settled. The molecular evidence is stirring the controversy even more.

Viral Diversity

Viruses are 100–1,000 times smaller than bacteria and are composed of DNA or RNA molecules wrapped in a protein coat. Since they are not enclosed in a membrane like all other organisms, they are described as acellular (without cells). Because they are incapable of an independent existence, all viruses parasitise other organisms. When they are not infecting a host, viruses behave as stable chemical molecules. Some argue that viruses lie between living things and chemicals. When infecting a host, viruses can multiply at a phenomenal rate. Viral diversity is currently impossible to assess but all groups—plants, animals, bacteria and fungi—contain them.

Viruses are important to understand because they cause diseases such as influenza, chickenpox and AIDS and remain a major challenge to medicine.

Suggested Reading

Brock, T.D., and Madigan, M.T., 1991.
Biology of microorganisms 6th ed.
Prentice Hall: New Jersey.

Margulis, L., Corliss, J.O., Melkonian, M. and Chapman, D.J., 1990.
Handbook of Protoctista.
Jones & Bartlett Publishers: Boston, Massachusetts.

Postgate, J.E., 1992.
Microbes and man 3rd ed.
Cambridge University Press: Cambridge.

Sagan, C. and Margulis, L., 1988.
Garden of microbial delights: a practical guide to the subvisible world.
Harcourt Brace Jovanovich: Boston, Massachusetts.

Woese, C.R., 1981.
Archaebacteria.
Scientific American June 1981.

Worldwide Patterns of Biodiversity

Most news reports about biodiversity concern the fate of tropical rainforests. While it is true that the warm and humid equatorial regions harbour the greatest concentrations of biodiversity, a myriad of species inhabit the Earth at other latitudes, both on the land and in the sea. Outside tropical rainforests there exist ecosystems just as fascinating, unique and important. In a few cases they also approach similar levels of biodiversity.

Life On Land

A good way to begin understanding the patterns of biodiversity on Earth is to look at the major types of vegetation, known as biomes. Their boundaries tend to run east–west, reflecting the change in climate from the poles to the equator. Tundra vegetation—both Arctic and Alpine—comprising lichens, mosses, sedges and dwarf trees, occurs in areas of extremely low temperature and rainfall. As temperature and rainfall increase towards the equator, the dominant vegetation changes from evergreen needle-leaved conifers, through temperate forests, such as *Eucalyptus,* to the evergreen broadleaved trees of tropical rainforests. Deserts occur in low rainfall areas across a range of temperatures. A slightly higher rainfall transforms them into grasslands. With good but seasonal rain, scattered trees occur throughout the grassland in the savanna biome. Regular fires under these conditions leave only fire-resistant shrubs—the chaparral biome. The east–west orientation of biome boundaries can be altered by mountain ranges that run north–south, such as the Rockies, Andes and the Great Dividing Range, as well as by other factors that influence climate.

Although biodiversity levels peak in the tropical latitudes, many species occur in the harshest environments. In spite of Antarctic temperatures reaching as low as –70°C, these Emperor Penguins maintain large colonies throughout the bitter winters. They maintain body heat by forming huddles and regularly rotating the birds on the outside into the warmer centre.

The Effects of Altitude and Latitude

In general, moving from the equator to the poles,

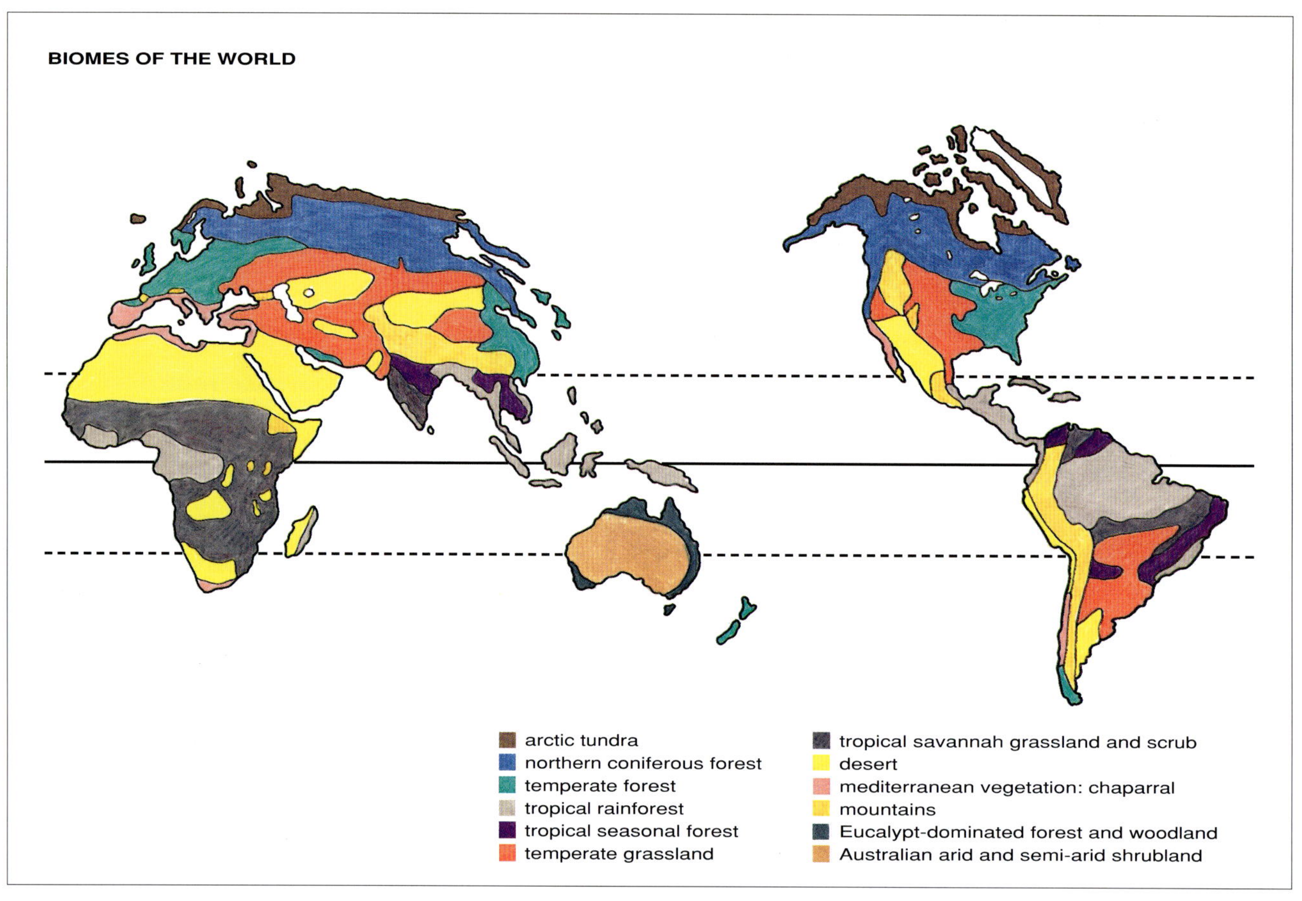

or from lower to higher altitudes, the climate increases in severity. Consequently, as a rule, the number of species declines. Tropical land areas, although they cover only seven per cent of the surface of the Earth, contain the largest number of terrestrial species—almost half the plants, nearly a third of the birds, half of all vertebrates and probably 90 per cent of arthropods. The number of species known near the poles or at high altitudes is much smaller.

Vegetation patterns on land are accompanied by large changes in diversity. This can be clearly seen on a worldwide scale: one tenth of a hectare of tropical rainforest often harbours over 150 tree species but the same area in a North American temperate forest has only 44 species, while there are only five species in the coniferous forests of Canada.

There are similar trends in animal diversity but as the total number of species at any site has not been measured, our knowledge of the changes must be based on just a few groups. Species of breeding birds, for example, increase from 22 to 1,823 between northern Alaska and the equator in Central America. Lizard and ant diversity show similar patterns.

While changes relating to climate take place over thousands of kilometres of latitude, similar marked differences occur in only a few hundred metres of altitude. Venturing up mountains from sea level, the number of species drops substantially over relatively short distances. For example, there are 320 bird species in Papua New Guinea living at about 250 metres elevation but this number drops to a single species above 4,000 metres. Mosses decrease from 182 species at 3,000 metres to three at 4,500 metres in these mountains. In Tasmania, only 88 species of geometrid moths occur in the highlands (above 900 metres) compared with 229 species in the lowlands.

Biodiversity Beyond Rainforests

Much attention has been focused on the destruction of tropical rainforests because they harbour so many species. However, many other ecosystems are equally or more threatened but receive little or no media coverage.

The coastal zones of the world cover eight per cent of the surface of the Earth. This is greater than the size of Africa. These zones include continental shelves, coral reefs, kelp 'forests', mangrove swamps, seagrass beds and intertidal marine platforms, most of which harbour exceptionally high levels of biodiversity. Unfortunately the proximity of coastal zones to very large human populations, as well as their importance to shipping have meant major threats to this biodiversity from urban and industrial pollutants, oils spills and, more recently, invasion by non-native species released with the ballast water of ships.

The great ecologist, Dan Janzen, recently wrote: 'The rainforest is not the most threatened of the major forest types. The dry tropical forests hold this honour. The story is the same for the dry tropical regions of Australia, South–East Asia, Africa and major parts of South America. The cause of severe habitat loss is straightforward. Dry forest is easily cleared with fire, and woody regeneration in fields or pastures is easily suppressed with fire. Furthermore, fire does not always stay where you put it and many areas are unintentionally cleared.'

Some dry tropical forests harbour as many species as wet tropical forests. However, worldwide, these forests are being converted to relatively species-poor grasslands by unnatural fire regimes and overgrazing. Frequently, key species are lost so that even when some trees are left, the forest is doomed. About this Dan Janzen wrote: 'These organisms now stand on a trashed agro-scape and will die without replacement. They are the living dead—all physiologically alive but can be regarded as dead as if they were already lying in the litter. If they flower, they lack pollinators and fail to set seed. If they set seed, the seeds lack dispersal agents. If they disperse, the right conditions for germination, growth and development are long gone.'

Coolibah trees are part of the plant biodiversity of the arid Pilbara district of Western Australia.

Other ecosystems with high diversity or with diversity composed of many rare species include those on mountain tops, oceanic islands, and semi-arid and arid areas. These tend to be places where the climate is harsh and many species are specialised for survival under these particular conditions and nowhere else. Often living conditions are so severe that populations are never large and species never common. Nevertheless, once the appropriate adaptations have evolved, some kinds of organisms flourish and become very diverse. Examples include the lizards and termites of the Australian outback, herbaceous flowering plants in alpine and sub-alpine meadows and the birds of remote islands.

Threats to tropical rainforest may have received the most publicity but biodiversity is in decline in many other kinds of ecosystems.

The general picture in mainland Australia is more complex because, in addition to the altitude–latitude pattern, the number of species also decreases from the humid north and east coasts to the arid centre.

Biodiversity of Islands

The number of species on an island generally depends less on its latitude than on how easily species can get to it (immigration) and whether or not they can gain a permanent foothold and thus avoid extinction. The immigration rate is influenced by distance from a supply of migrants, usually at its highest on the mainland. The rate of extinction is affected by island area: the bigger the island the more likely that there are suitable habitats for the arriving immigrants. Mobility clearly plays a part in the survival of new arrivals but even those with the greatest stamina find distance is important.

In general, the bigger the island, the more species it can accommodate. A good illustration is the number of birds inhabiting different islands at roughly the same latitude and distance from mainland New Guinea. This increases from 15–20 species in tiny islands of only one square kilometre in area to 120 in islands about the size of Kakadu National Park (17,000 square kilometres).

The loss of biodiversity from islands has been massive. Perhaps the best-known losses are of birds, including flightless geese and ibises in Hawaii, giant moas in New Zealand, the Dodo of Mauritius, and on the tiny island of St Helena in the South Atlantic where only one out of the five native bird species has survived. A recent survey of 13 oceanic islands has shown that between 51 per cent and a staggering 96 per cent of endemic flowering plant species are rare, endangered or already extinct. On some islands, such as Lord Howe and Norfolk, only one or two native species can be excluded from these categories, and on others all native flowering plant species are in trouble.

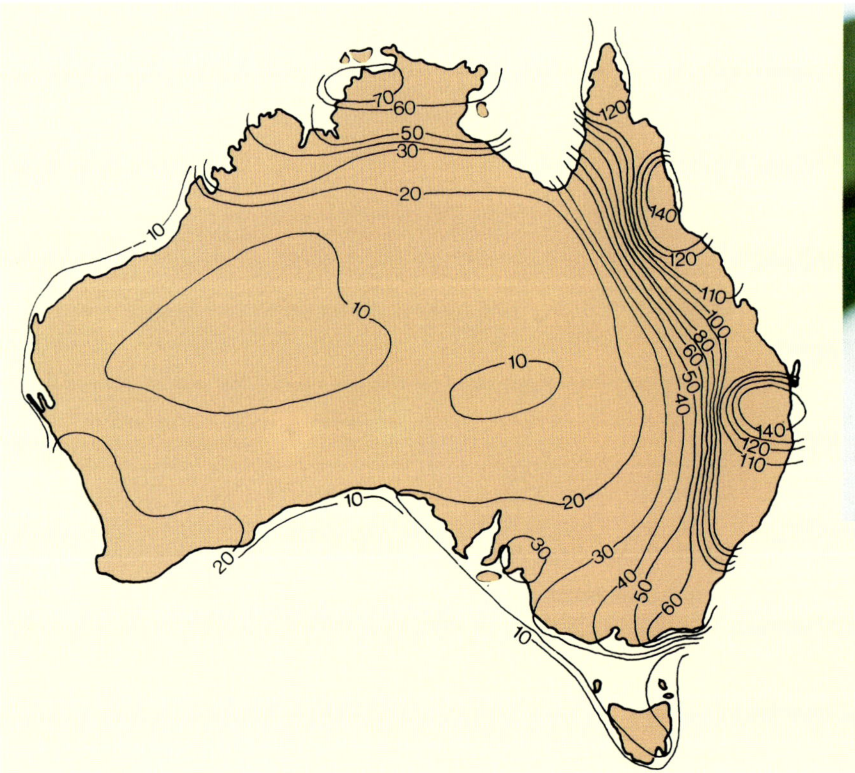

The pattern for butterfly distribution in Australia is similar for many other organisms. Peak diversity of over 140 butterfly species occurs along the Great Dividing Range and declines to fewer than 10 species towards the arid centre. The darker the shading, the greater the number of species. **ABOVE** *A birdwing butterfly from the high-diversity eastern coast of Queensland.*

However, this trend is not restricted to islands of land surrounded by water. The same principles apply equally well to islands of remnant vegetation surrounded by farmland, isolated mountain tops, lakes and, especially, national parks and nature reserves surrounded by agricultural land, urban developments or cities (see Chapter 9).

Islands are particularly vulnerable to ecological damage. A third of the plant species on Lord Howe Island, off the coast of New South Wales, are considered endangered.

Biodiversity in Mediterranean Climates

Another exception to the general equator-to-pole gradient in biodiversity occurs in regions with hot dry summers and cool wet winters—the Mediterranean climates. Although distant from their namesake, these regions include California, the Cape Region of South Africa and parts of south-eastern and south-western Australia. They have more plant species than do the other temperate regions, in fact their numbers approach those of tropical forests. For example, a tenth of a hectare in humid tropical rainforest in Queensland harbours 141 flowering plant species but the Cape Region of South Africa supports up to 130 species and in eastern Australian eucalypt woodland, there can be as many as 93 species in the same-size area.

Life in Fresh Water

Freshwater biodiversity follows similar geographic patterns to those on land. The number of stream-inhabiting insects is greater in the tropics than in mid-latitudes and, within tropical stream systems, greater at lower altitudes. In Australia, dragonflies and damselflies decrease from north to south along the east coast: from 116 species in south-eastern Queensland, through 85 species in south-eastern New South Wales, to 65 species in Victoria and 26 species in Tasmania.

Compared with land and ocean systems, freshwater habitats—rivers and lakes—are relatively isolated and the dispersal of organisms is prevented by land and marine barriers. This has led to a high degree of endemism (the restriction of a species to a unique location). Consequently regional biodiversity can be high even if local biodiversity is low.

Endemism is particularly high in lakes—90 per cent of the fish species from the rift lakes of Africa and from Lake Baikal in central Asia occur only in those lakes. Even in river systems, fish endemism can be high. In Australia, many of the freshwater fish species are restricted to one or two river systems. For example, although the Spangled Perch, *Leiopotherapon unicolor,* lives in all mainland Australian river systems, other members of the same genus are highly restricted. A related fish, the Black Grunter, *L. macrolepis,* is found only in the

Prince Regent and Roe Rivers, which drain into the Timor Sea. Another, the Fortescue Grunter, *L. aheneus,* occurs only in the Ashburton, Fortescue and Robe Rivers, which drain into the Indian Ocean. This is a common pattern.

about 800 at the equator. The number of reef-forming coral species on the Great Barrier Reef exhibits strong geographic variation rising from below 10 in the south to over 50 near Cape York.

Superimposed on this general geographic pattern is another pattern associated with depth. In the shallowest—the illuminated—zone, there are about 8,000 species of seaweed worldwide and 50 species of seagrass (marine flowering plants). The number of species decreases as the water gets deeper and darker.

Local versus Regional Biodiversity

Where the variety of habitats remains relatively constant as a result of unchanging climate across a region, regional biodiversity closely resembles local biodiversity. In other words, estimates of biodiversity in an area the size of two football fields give reasonable estimates of regional biodiversity, such as an area the size of New South Wales. On the other hand, in geographically varied areas, with differences in climate and a large variety of habitats, local biodiversity is considerably less than the biodiversity of the region as a whole. Here, local estimates are no guide to overall regional biodiversity.

Life at Sea

Compared with land environments, the marine environment is less accessible and nowhere near as well researched. From the little we do know, marine diversity shows the same kind of general trends with latitude as are found on land. For example, on coastal land, mangrove species in Australia decline from north to south along the eastern seaboard: from 30 in northern Queensland to none in Tasmania. In the adjacent ocean, seagrass species follow the same pattern: from 12 species in the north to between one and four in the south. Seaweeds are even more interesting as the number of species, while generally increasing from the poles to the tropics (from 168 in the Canadian Arctic to 750 in tropical West Africa), reaches a peak (1,100) in southern Australia, a Mediterranean climatic region. Marine bivalve molluscs—shellfish with two hinged shells—increase from between 66 and 100 species near the poles to

Although far less well known, the oceans are of great importance to biodiversity simply because they are so vast. They cover 71 per cent of the surface of the Earth—361 million square kilometres—and the entire water volume is a habitable environment, with different creatures living at each and every level. Considering that maximum depth reaches roughly 11 kilometres in the Mariana

Although far less well known, the biodiversity of oceans seems to be extremely high. Floating on or near the surface are many groups of animals. This nudibranch or sea slug is feeding on the tentacles of a bluebottle jellyfish.

Trench, in the Pacific, the potential for ocean biodiversity is enormous.

Ocean Environments

Ocean surface waters are divided into broad life zones that reflect climatic variation. They have descriptive names such as Arctic, Antarctic, Northern and Southern Hemisphere Cold Temperate, Northern and Southern Hemisphere Warm Temperate, and Tropical. The boundaries of these regions follow surface temperature zones. In coastal regions the change from one region to the next is often accompanied by marked changes in flora and fauna.

At the surface of the ocean, organisms are exposed to full sunlight. Surface temperatures vary as they do on land but much less so: from about –1°C in Arctic and Antarctic waters to more than 20°C in equatorial waters. Temperature is linked to variations in the amount of light. There is little variation in light throughout the year at the equator, but near the poles days reach their peak in summer and decline in winter to perpetual night.

The physical nature of the oceans changes dramatically with depth and is important in determining the types of organisms that utilise the different zones. The upper, sunlit zone is the only habitat where phytoplankton—the primary producers—can exist since they require light energy for photosynthesis. Over three-quarters of the ocean floor—77 per cent—is found in the abyssal province. Eight per cent of the ocean is continental shelf and the remainder occupies the transition between the two. Oceanic trenches make up less

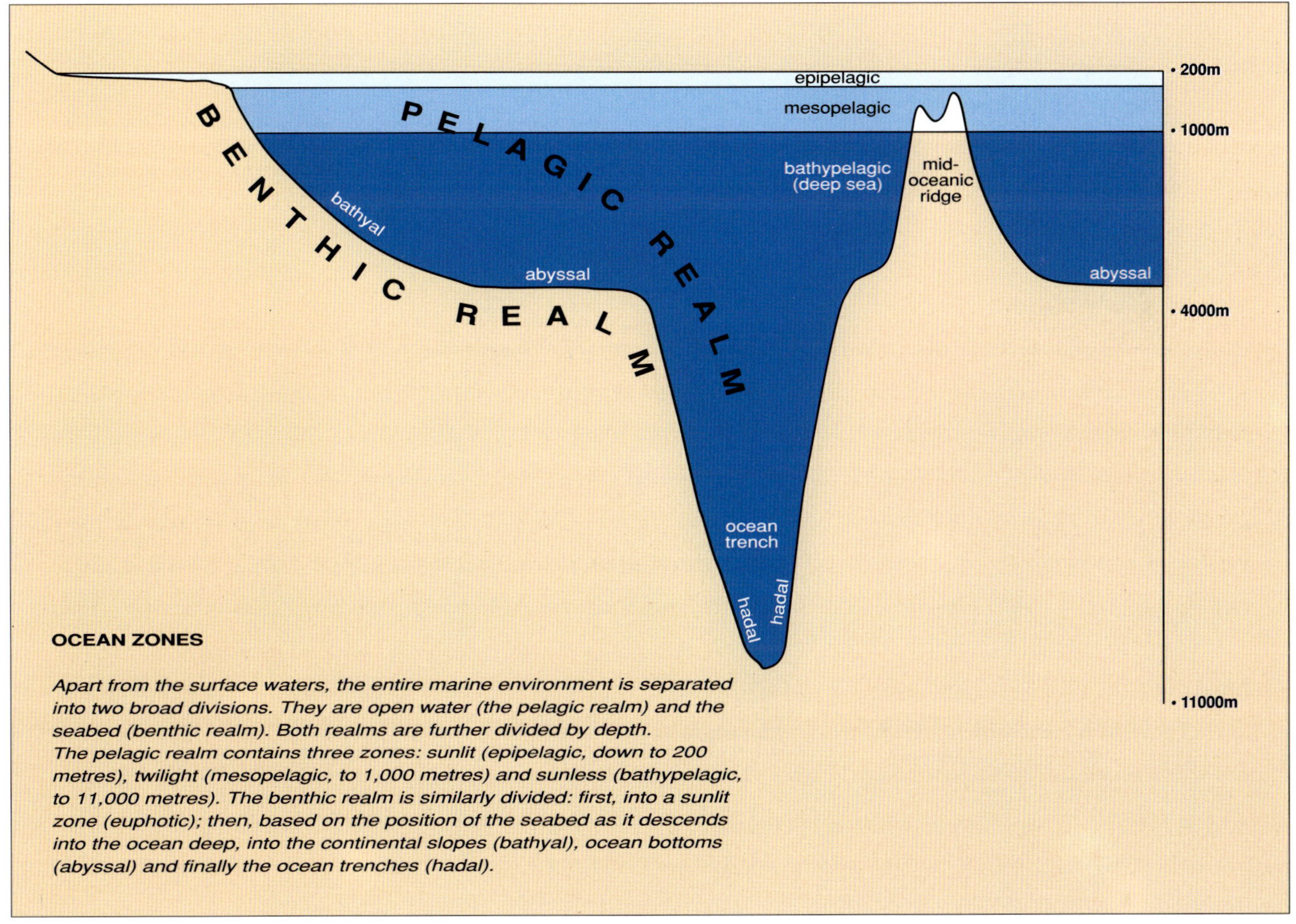

OCEAN ZONES

Apart from the surface waters, the entire marine environment is separated into two broad divisions. They are open water (the pelagic realm) and the seabed (benthic realm). Both realms are further divided by depth. The pelagic realm contains three zones: sunlit (epipelagic, down to 200 metres), twilight (mesopelagic, to 1,000 metres) and sunless (bathypelagic, to 11,000 metres). The benthic realm is similarly divided: first, into a sunlit zone (euphotic); then, based on the position of the seabed as it descends into the ocean deep, into the continental slopes (bathyal), ocean bottoms (abyssal) and finally the ocean trenches (hadal).

than one per cent of the world's benthic marine environment.

Pressure increases with depth. At the surface, pressure varies only by a small amount around one standard atmosphere, just as it does on land, depending on the movement of high and low atmospheric pressure cells across the surface of the Earth. However, water is heavy and, at about 10 metres depth, the pressure is twice that of the surface. At the greatest depths the pressure increases until the column of water can exert over 1,000 atmospheres of pressure on living organisms.

In contrast, light, oxygen and temperature decline with depth. At the bottom of the illuminated zone, light levels are too low for photosynthesis to occur. This boundary varies with the quantity of materials suspended in the water. Enough light can penetrate to 200 metres in clear ocean waters but does not pass beyond much shallower depths in turbid coastal and estuarine waters. This limits the distribution of plant species.

The relationships between oxygen, temperature and depth are complex. Surface waters are almost completely saturated with oxygen; they can hold no more. The amount of oxygen that seawater can hold depends on temperature, so there is more—perhaps 50 per cent more—oxygen in the cool polar surface waters than in the warmer waters near the equator. Oxygen values decline with depth.

Water temperature at abyssal depths is one to four degrees Celsius but can fall to zero at the bottom of ocean trenches. On the other hand, very high temperatures (325°C) occur near thermal vents on the ocean floor (see box in Chapter 5).

The number of species living on the seabed, such as molluscs and shrimps, rises gradually with increasing depth to a maximum at between 2,000 and 4,000 metres but declines at greater depths. These numbers are influenced by two factors: availability of food and penetration of light.

Marine ecosystems have the highest animal biodiversity in terms of phyla—28 compared with 14 in fresh water and only 11 on land. Biodiversity is particularly high on coral reefs. For example,

This spectacular stomatopod or mantis shrimp was recently discovered in the Coral Sea. It lives in near-darkness at a depth of 1000 metres and preys on other animals with its powerful forelimbs.

What are Phyla?

At the most general level, organisms are assigned to different kingdoms (see box on page 12). The next level of classification within a kingdom is the phylum (plural, phyla). Phyla are defined by some major feature or combination of major features that makes their 'blueprint' different from all others. At this level of classification many features that may seem distinctive are irrelevant. For example, although a group of animals may have external coverings as different as hair, feathers, slimy skin or scales, their blueprint is unique as they almost all have structures built around a backbone with a massive nerve running along it. This phylum is the Chordata and includes fish, frogs, reptiles, birds and mammals (including humans). Currently 92 phyla are recognised across all kingdoms, that is, 92 highly distinctive blueprints.

It would be natural to think that exploration of the world was so far advanced that all phyla have been discovered. Not so—as recently as 1983, a new phylum was discovered and called the Loricifera. These organisms are tiny animals adapted to live in the spaces between grains of sand to which they cling by means of hooks on their heads. People working with biodiversity will not be too surprised to find there are more undiscovered phyla lurking in the microscopic jungles of the world.

In the oceans, biodiversity patterns alter with depth as well as latitude. Coastal zones can exhibit high levels of biodiversity, such as in seaweed 'forests', which occur only in the illuminated zone.

great depths—below 6,000 metres—489 species, ranging from single-celled amoebae to fish, were found. Endemism is high at such great depths: of the 489 species, over two-thirds occurred exclusively around 6,000 metres. Such endemism appears to decrease in shallower waters. For example, of the fish species found above 2,000 metres in the eastern and western Atlantic Ocean, 40 per cent occur in both areas. However, below 4,000 metres there is little overlap.

The bottom-living fauna of Australian waters may be even richer than elsewhere—recent samples of just 10 square metres in Bass Strait yielded more than 800 species. Moreover, it has been suggested that, as the oceans of the southern hemisphere are much more extensive than those of the north, new calculations of ocean biodiversity that include this factor will greatly increase estimates of the total number of species inhabiting the marine environment.

over its full range, the Great Barrier Reef and its cays support 1,500 species of fish, 252 species of nesting birds, five species of turtles and various species of whales and dolphins. These animals only represent a single phylum. Imagine the diversity within the other 27 phyla here.

Deep-sea communities, too, can be very diverse because they have high endemism: nearly 800 species from 14 phyla have been collected off the coast of New Jersey alone. Even at surprisingly

How the Patterns Were Established

These patterns of distribution of life over the surface of the Earth reflect its turbulent geological and climatic history. Looking at an atlas, it is difficult to imagine that the present position of the continents is not fixed. But close inspection of their shapes suggests that the continents once fitted together like the pieces of a jigsaw puzzle. The southern coastline of Australia mirrors parts of

CONTINENTAL DRIFT

Movement of the continents since the break up of Pangaea.

200 million years

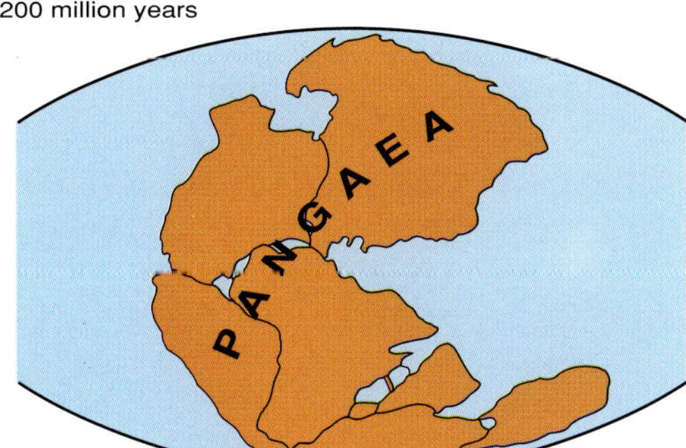

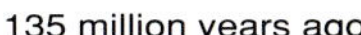

135 million years ago

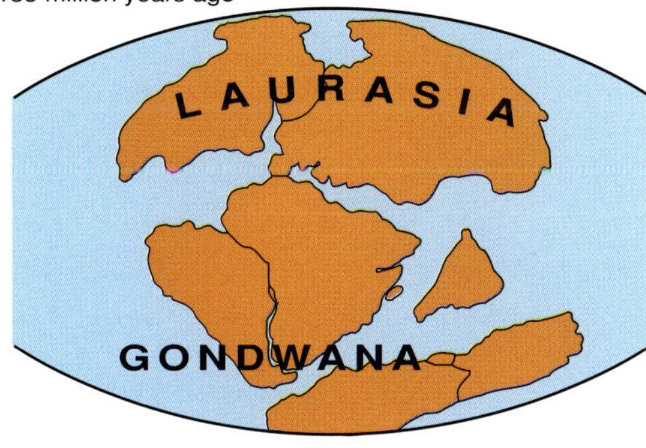

65 million years ago

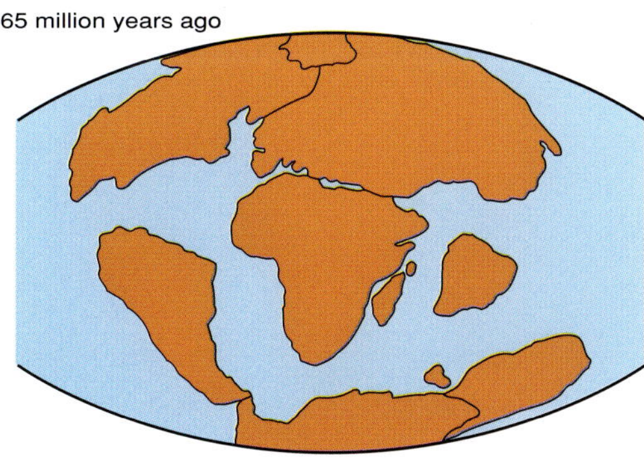

Present

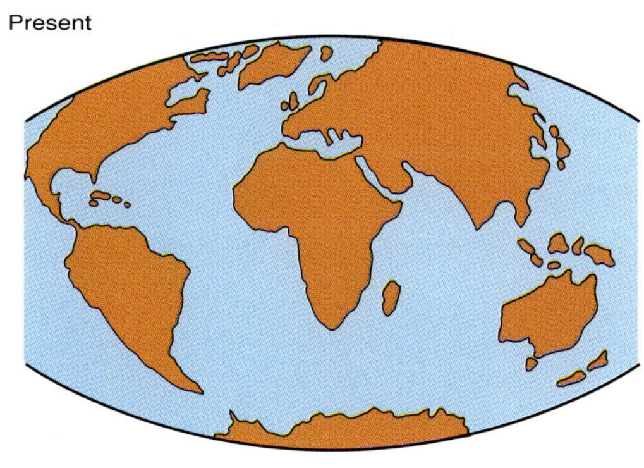

Antarctica; the east coast of South America would fit snugly into the west coast of Africa, and North America and Greenland would once have nuzzled the west coast of Europe. This is no coincidence: over 300 million years ago the continents were joined together in a single supercontinent called Pangaea. Over vast time spans, the continents have shifted over the globe, a process known as continental drift.

Currents in the molten rock beneath the crust of the Earth move the giant crustal plates, which support the continental land masses. This process is like the continual formation and break up of Antarctic ocean icefields. The icefloes move on the surface of the ocean to form extensive, hilly pack-ice during winter, which breaks up in spring leaving open bodies of water between smaller rafts of ice. In a similar way, the continents slowly split and became surrounded by oceans.

The forces that drove the movements of continents in the past are operating just as much today: Australia is moving northwards six centimetres per year, about the rate fingernails grow.

The isolation of Australia, accompanied by unusual climatic changes, contributed to its unique biodiversity, which is discussed separately in the next chapter.

Suggested Reading

Colinvaux, P., 1993.
Ecology. Wiley: New York.

Furley, P.A., Newley, W.W., 1983.
Geography of the biosphere. An introduction to the nature, distribution and evolution of the world's life zones.
Butterworths: Sydney.

Goodall, D.W., 1977.
Ecosystems of the world (15 volumes).
Elsevier Scientific Publishing Co.: Amsterdam.

Janzen, D., 1988.
Tropical dry forests: the most endangered major tropical ecosystem.
In: **Biodiversity**, E.O. Wilson (Ed.) pp. 130-137.
National Academy Press: Washington D.C.

Ricklefs, R.E., 1990.
The economy of nature.
W.H. Freeman & Co.: New York.

Reid, W.V. and Miller, K.R., 1989.
Keeping options alive: the scientific basis for conserving biodiversity.
World Resources Institute: Washington, D.C.

Australian Biodiversity

Australia supports many species of unusual plants, animals and micro-organisms, and the natural communities in which they are found, which are not present anywhere else on Earth. The sheer numbers and variety of species have led Australia to be classed as one of only 12 megadiverse nations. The wealth of life in Australia is diverse in almost every aspect—in its origins, in its variety of organisms and in the extraordinary radiations of some groups, as well as in its unique ecosystems. In broad terms, the patterns of biodiversity in Australia are similar to those in the rest of the world. However, millions of years of isolation from other continents and the climatic history of Australia have combined to create distinctive life zones and a unique array of living things.

The distribution of life today reflects the turbulent geological history of the Earth. Ancestors of Nothofagus *trees, like this Deciduous Beech from Tasmania, were present across Gondwana. Today modern forms are found in Australia, South America, and New Zealand.*

Origins of Australian Biodiversity

By the time the single landmass of Pangaea began to break up—about 160 million years ago—the antecedents of all the major groups of plants and animals had already evolved. Algae, mosses, ferns, cycads and conifers, and primitive flowering plants, cloaked the land. Worms, insects and other arthropods, amphibians, reptiles (including dinosaurs), and even early mammals, fed upon the vegetation and each other. Bacteria and fungi carried out their vital roles in decomposition and recycling.

Pangaea slowly split into two supercontinents, separated by an intercontinental ocean, the Tethys Sea. To the north lay Laurasia, comprising what is now North America, Europe and Northern Asia. All the southern landmasses were joined—Antarctica, South America, Africa, Madagascar, Arabia, India and Australia with New Guinea. Together they formed Gondwana.

Laurasia and Gondwana each inherited its share of Pangaean biodiversity, which is why the flora and fauna of the two hemispheres, although distinctive, comprise the same basic groups. Following the break up of Pangaea, however, the land organisms on the two supercontinents evolved independently and, as a consequence, different groups diversified and became dominant. We can still witness many echoes of this differentiation:

• The ancient velvet worms (Peripatus) left only fossils in North America, Europe and Asia but persisted in Gondwana. Today they are found only in South America, Africa, Australasia and a small Gondwanan fragment that is now part of South–East Asia.

• A large family of plants, the Proteaceae, originated in Gondwana. Today it is found in South America, India and South Africa, where it is known for its *Proteas*, and in Australia, where it is represented by the widespread *Banksias, Hakeas* and *Grevilleas.*

• There is also an interesting contrast between the mammal faunas of the northern, Laurasian, and southern, Gondwanan, continents. While marsupial fossils can be found in the Laurasian landmasses, placental mammals—such as antelopes, bears, cats and primates—gained their stronghold there. Conversely, the marsupials gained supremacy in the south. It had been presumed that marsupials were somehow inferior and only thrived in Australia because they did not have to compete with placental mammals. However, the recent discovery of ancient placental fossils in Australia does not support this view. Neither does the modern fauna of South America, where a rich marsupial diversity persists despite massive placental invasions from North America.

After Pangaea divided, the crust of the Earth continued its slow but relentless movements and Gondwana itself began to break up. India began to move northwards 118 million years ago, where, much later, its massive collision with Asia created the Himalayan mountains. Africa, Australia and South America slowly separated from the dwindling remnants of Gondwana. Signs of the initial rifting between Australia and Antarctica about 170 million years ago can be seen today in the massive dolerite cliffs in Tasmania, a spectacular example being at Cape Raoul on the Tasman Peninsula. This dolerite matches the dolerite of the same age in Antarctica.

A member of the Gondwanan family Proteaceae, Banksia coccinea.

During the lengthy process of separation, a

Ancient Soils

Over the last 20–30 million years, Australia has been through a period of remarkable geological stability. Major earthquakes have been rare and volcanic activity restricted to a narrow belt down the eastern and southern seaboard. Viewed from a human perspective this may seem to be a boon—particularly when compared to our northern neighbours on the western rim of the Pacific. However, this very stability has imposed a stress on vegetation that has moulded the evolution of much of our plant life. Mountain uplift followed by erosion, together with volcanic lavas and ash, bring new nutrients to the soil. In their absence, minerals essential to plant growth have been slowly leached from Australian soils.

Rainforests are relatively resilient to such conditions. Humid microclimates promote rapid decay of dead organic matter, and even limited nutrients can be recycled through the system. Provided with adequate nutrients and water, plants can afford to produce broad, soft leaves that are easily damaged, shed and replaced by new growth. With increasing aridity and unpredictability of the climate, the cycle became fragile over much of the continent. On the edges of the dwindling rainforests many different types of flowering plants evolved a frugal adaptation that is now a feature of much of our bushland. Leaves became smaller, thicker and leathery, and much longer lasting. There is evidence that plants with these sclerophyll (literally 'hard-leaf') characteristics were evolving in nutrient-deficient swamplands 20 million years ago—long before Australia became the dry continent. However, these adaptations to impoverished soils had a vital and beneficial spin-off. Sclerophyll features also permitted plants to resist the stress of low water availability and to colonise the expanding dry lands.

truly Earth-shattering event occurred. Sixty-five million years ago, mass extinctions decimated the flora and fauna of the world. While numerous theories have been postulated, the most widely accepted is that a bolide (a meteorite several kilometres in diameter) smashed into the Earth, throwing up a thick dust cloud, blanketing the planet and drastically reducing the sunlight reaching the surface and, therefore, the photosynthetic output of plants. The casualties of this holocaust included the majority of the giant tree ferns, conifers and their relatives and the dinosaurs. However, many ecological niches became available for exploitation by the modern mammals and flowering plants.

The last connection between Australia and Antarctica, to the south of Tasmania, finally broke about 40 million years ago. At this time, the Australian fauna contained a large inventory of invertebrates. Also frogs, fish and turtles inhabited the creeks and pools; birds and bats flew through and above the canopy; and the ancestors of the Emu and cassowaries picked over the forest floor. Snakes and lizards were present. Among the great variety of marsupials, the only types that would be familiar to us today were primitive bandicoots and carnivorous dasyurids, resembling present-day quolls.

Although Gondwana straddled the South Pole, the climate at that time was much warmer and wetter than it is there today. Rainforest trees, such as plum pines (*Podocarpus*) and southern beech trees (*Nothofagus*) covered Gondwana. *Nothofagus* and *Podocarpus* species are still found in Australia.

The break up of Gondwana had a direct impact on the climatic patterns of the world. The northerly movement of Australia and the final severance of South America from Gondwana opened a new sea passage, enabling ocean currents to circulate around Antarctica. Thus oceanic currents were no longer directed around the Australian

After Antarctica, Australia is the driest continent on Earth, yet it has no deserts devoid of vegetation like the Sahara. While much of the continent has been colonised by eucalypts, the wattles dominate the drier regions.

coastline to mingle with warm equatorial waters as they had done in the past. The circulation of the cool polar waters in the south created a 'refrigeration effect' around Antarctica, eventually creating the much stronger gradients in temperature between the pole and the equator that exist today.

A consequence of these changes was that, as Australia moved north, its climate became drier. The lush *Nothofagus* forests became restricted to the southern half of the continent and relatively few of the ferns or the conifers and their relatives persisted. As the Gondwanan forests dried, new selection pressures favoured plants that could withstand the new conditions of low water and nutrient availability.

After separating from Antarctica, Australia travelled nearly 30° of latitude. This period is sometimes referred to as the time of the Great Southern Ark. Although its isolation from other continents during this northerly drift was undoubtedly important in the biological history of Australia, the degree of insulation from world biodiversity can be over-emphasised. Many marine and intertidal organisms never lost the potential for intercontinental exchange. Similarly, we are only now realising how easily some bacterial, fungal and algal spores, small seeds and invertebrates, such as spiders and flying insects, can drift on the air currents of the world.

About 25 million years ago, larger organisms began to cross the narrowing straits between South–East Asia and Australia. For example, 20 million years ago, Australia had a bat fauna as abundant as, and remarkably similar to, that which coexisted in Europe. About this time, two new players took on increasingly important roles on the botanical scene—the eucalypts (members of the Gondwanan family Myrtaceae), which much later came to dominate the moister regions of Australia, and the wattles or *Acacias*, which now hold sway in the dry interior. In the forests of the time, most types of possum had already evolved, together with many that are now extinct. The kangaroos (macropods) had diversified but were very different in their habits to their predominantly grass-grazing modern counterparts. Some browsed on the leaves and fruits of shrubs and trees while others developed the teeth and jaws of carnivores.

Over the last few million years, as Australia approached its present position with respect to Asia, increasing dryness and unpredictability in the climate finally opened up the forests. Rainforest retreated to the more mountainous and wetter areas and *Eucalyptus* forests developed an understorey in which grasses were often dominant. Spinifex grasses with *Acacia* shrubs became a key feature of the semi-arid environment. The stage was set for the radiation of the insects, reptiles, birds and mammals of the interior—including the modern kangaroos. Ancestors of our native rodents entered the Australian continent in several invasions and were abundant in the drier areas by four million years ago.

This was not the end of the massive forces that moulded the biodiversity of Australia. Beginning nearly three million years ago the ice caps at the poles advanced and retreated causing prolonged periods of severe cooling (glaciations) with intervening warmer periods (interglacials). The resulting sea-level changes affected much of the world as bridges were formed between lands previously separated by ocean, permitting migration. During periods of low sea level both New Guinea and Tasmania were connected to the Australian mainland for varying periods. However, actual encroachment by ice affected only southern Australia, especially Tasmania, most recently between 30,000 and 12,000 years ago, determining, in particular, the distribution of the cold-adapted alpine and sub-alpine floras.

The most recent event, and the most dramatic in terms of the rate of change it induced, was the advent of our own species. Aboriginal people arriving over 40,000 years ago did not bring many new species with them. Their deliberate burning of tracts of bush in their husbandry of the land did, however, alter much of the Australian vegetation. In contrast, Europeans attempted to introduce many components of northern hemisphere ecosystems. Vast areas of land were cleared and irrigated to farm cereals, vegetables, fruits and livestock for a growing population. Out of sentimentality, and for comfort in an apparently alien land, came sparrows, blackbirds, domestic cats and all the trees and flowers of an English garden. For sport came rabbits, foxes and deer. The effects on native ecosystems are discussed in Chapter 9.

Ancient, Odd and Unique

With such a unique and varied history, it is not surprising that the fauna and flora of Australia contain many organisms not found anywhere else on Earth. Some represent life forms long extinct on the other continents. Others are unique evolutionary adaptations to our often harsh and unpredictable climate. Some are well known, some are known only to experts and many are still being discovered. All are a part of our national heritage. It is a reflection of this rich diversity that only a tiny proportion can be included in this chapter.

In a *Guinness Book of Records*-style list, the entry for Australian biota would be impressive. For sheer longevity, the stromatolites living in the salty waters of Shark Bay in Western Australia must take a prize. Formed by primitive photosynthetic bacteria, these structures closely resemble their three-and-a-half billion-year-old fossil relatives found inland. The corals of the Great Barrier Reef hold the record for bulk—as Australia moved northwards into warmer latitudes, these tiny animals progressively laid down a structure unsurpassed in size even by the works of humans. We can also boast the tallest flowering plant in the world, *Eucalyptus regnans,* the Mountain Ash of Victoria and Tasmania, which can reach to a height of 100 metres; and the Giant Gippsland Earthworm, which grows to an almost unbelievable three metres in length.

Organisms do not need to be big or ancient to be special. For example, one species of ant has been of great interest to the worldwide community of entomologists. In 1931, two unusual specimens were found among a collection of insects from coastal heathland east of Esperance in Western Australia. They appeared identical to 60-million-year-old fossil ants. Forty-six years of national and international expeditions finally turned up living colonies, this time on the Eyre Peninsula, and *Nothomyrmecia macrops* was recorded as the most primitive living ant—a living fossil. Other insects show associations with plants that must date back to their common Gondwanan origins. The plum pines and the southern beeches are hosts to their own unique species of primitive sap-sucking aphids—ancient enemies of ancient plants. In the

In a remarkable adaptation to arid conditions, desert burrowing frogs bury themselves in drying mud and remain in a state of suspended animation, sometimes for years, until rains soften the soil and allow them to emerge.

case of the southern beeches, the relationship has a further twist: the aphids are, in turn, parasitised by a very primitive wasp!

The crustaceans, although predominantly aquatic, have found ways to exploit terrestrial Australian environments. Slaters are abundant in the moist forests but more adventurous species have adapted to life in the cracks in soil and under stones, even in semi-arid regions. Some crustaceans appear to have gone to the other extreme—they *require* desiccation. For example, the eggs of shield shrimps must be thoroughly dried before they can develop. For years they are blown around in the dust until the rains come and allow them to hatch. Ephemeral puddles in central Australia teem with many thousands of these four-centimetre-long curiosities. Within a few days their life-cycle is completed, and a new batch of eggs is dispersed on the wind as the puddles dry. An even more improbable arid Australian inhabitant is the desert crab, *Holthusiana transversa,* which lives in clay burrows near sites of sporadic flooding. Unlike most crabs, this animal does not require permanent water, breathes through lungs and uses its gills only when it takes to water to feed, moult or mate.

Related to the crabs are the freshwater crayfish, often called yabbies. For such a dry continent, it is surprising that Australia supports over 100 species—surpassed only by North America with 300. These include the largest in the world, *Astacopsis gouldi,* from Tasmania, weighing in at a massive three-and-a-half kilograms and 40 centimetres long; and the featherweight of the world, *Tenuibrachiuris glypticus,* from Queensland, at just a few grams and two-and-a-half centimetres in length. Some yabbies have taken to a burrowing life in the waterlogged soils of the Tasmanian mountains where they may share their burrows with ancient relatives, the mountain shrimps (syncarids). These represent a key group in the evolution of the prawns, crabs and yabbies. Their fossils date back over 180 million years but living forms are still to be found in the cool streams of the Victorian and Tasmanian highlands and, in the case of a few species, in the burrows of their yabby relatives. These examples illustrate that although Australia is a dry continent, its surface waters are havens for a great variety of unique and fascinating animals.

Australian Vertebrate Diversity

The vertebrate animals of Australia are at least as diverse as those of any other landmass. Among the freshwater fish are the targets of recreational anglers: the Murray Cod, barramundi, and the Saratoga of the tropics. Of particular interest to biologists are the fish that have evolved in highly restricted areas of the inland drainages. For example, the only known habitat for *Scaturiginichthys vermeilipinnis*, a member of the blue-eye family (pseudomugilids), is five shallow springs on a single pastoral property near Longreach in central western Queensland.

Perhaps the most unusual fish of all is the Queensland Lungfish, considered to be a living fossil intermediate between the true fishes and the amphibians. Lungfishes also survive in Africa and South America but *Neoceratodus forsteri* appears to be the most primitive of them all, resembling the fish that gave rise to the land vertebrates over 300 million years ago. It is now restricted to a very few rivers in south-eastern Queensland where it lives in deep waterholes and grows up to one-and-a-half metres in length.

The modern lungfishes represent only one tiny branch of the family tree coming from the ancient fish. Far more abundant and diverse are the descendants that became the amphibians. These in turn gave rise to reptiles, birds and mammals. The tailed amphibians (newts and salamanders) are not found in Australia but the frogs and toadlets are surprisingly well represented for such a dry continent. Some have adapted to life in the harsh interior—for example, the desert burrowing frogs. These frogs burrow more than 30 centimetres into the ooze of a drying billabong where they create a cocoon by compressing the mud around them and building a protective sheath of dead skin over their bodies. In a state of suspended animation, they can survive for several drought years, until the next rains soften the soil and allow them to emerge to mate and feed.

Other Australian frogs may live in more conventional habitats, such as the fast-flowing creeks of the Great Dividing Range. Even here some have

Thorny Devils are well adapted to arid regions. Their spines collect dew, which is transported into the mouth via a network of interconnected channels.

evolved unique habits. To avoid their offspring being washed away, the females of gastric brooding frogs gulp down their own freshly fertilised eggs, which then develop inside their stomachs. Not surprisingly, the mother has to suspend her usual stomach processes to avoid digesting her offspring. Giving 'birth' for these species is a form of regurgitation of self-sufficient froglets. It is alarming that no specimens of these frogs have been found for some years.

Among the great variety of snakes, lizards and turtles, perhaps the most fascinating are those that have adapted to the dry centre. The Thorny Devil, *Moloch horridus,* is named after a Semitic god for whom children were burnt in sacrifice. Fortunately it does not live up to its namesake, nor is it horrid. This gentle reptile shelters under spinifex clumps and offers little resistance when disturbed. The imposing spines, which cover its head and body, may help deter some predators such as snakes or birds of prey but their main function is for water collection. On cool desert nights, dew forming on the spines is carried by fine channels that run across its body and deliver water to its lips.

No account of Australian wildlife would be complete without mention of the only surviving egg-laying mammals in the world, the monotremes: the Platypus and echidnas. These animals retain some reptilian features of the skeleton and the habit of laying shelled eggs. However, most embryonic development occurs while the egg is retained in the reproductive tract of the mother. The eggs hatch shortly after laying and the young are nourished by milk—not via a teat but from patches of skin on the abdomen of the female. These animals are not, however, typical of the ancient monotremes as they are highly specialised as aquatic foragers and anteaters respectively. It is probably because of this extreme specialisation that these animals have survived to the present day.

Australian Plant Diversity

The northward migration of Australia took it into the warmer and drier latitudes, which greatly affected the vegetation. As the interior became hotter and drier, the rainforests retreated to the coastal zones. Those plants and animals that were adapted to cooler and moister conditions found a haven in south-eastern Australia and Tasmania. For some species this was a last ditch retreat because their particular requirements of temperature and moisture were found only in very small geographic areas. These areas are known as refugia and are particularly important in the Great Dividing Range where the mountainous terrain has provided a variety of such suitable environmental conditions.

Remnants of the Gondwanan rainforests persist: together with ferns and tree ferns, Australia

Eucalypt forests and woodlands dominate the moister areas of Australia and are even adapted to the snow country. Most of the 500 Eucalyptus *species have very restricted ranges.*

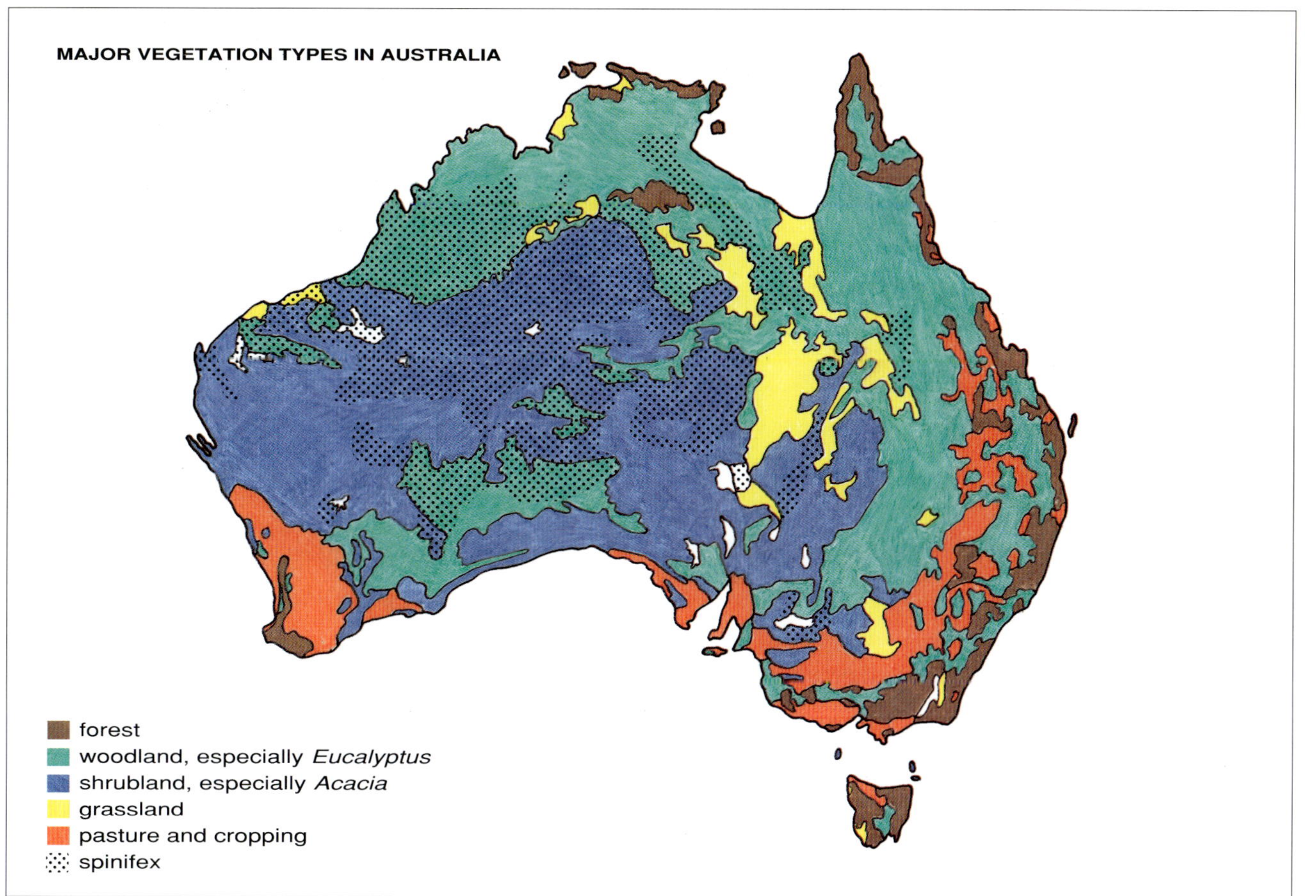

still possesses a diversity of cycads. Two hundred million years ago, these plants were widespread and numerous in the vegetation of the Earth. Now there are only 10 genera confined to the tropics and sub-tropics of central America, Asia, Africa and Australia. Of these, Australia possesses four, including the most southerly populations in the world, on the coast of New South Wales.

Australia remains home to some plant species that are considered to be the most primitive flowering plants. Some of them show remarkable traits that suggest how the flower evolved. For example, the pollen-producing organs, or stamens, are generally highly specialised with packets of pollen borne on thin stalks. However, some primitive flowering plants in Australia have stamens that look like petals or pale leaves, with the pollen packets clustered on the upper or lower surface. This trait shows that flowers have evolved from clusters of highly specialised leaves. The outermost are often brightly coloured to attract pollinators. These surround others that bear either pollen or ovules. Two genera with such traits are *Idiospermum* and *Austrobaileya,* both of which occur in northern Queensland and nowhere else: so unique are they that they are placed in their own separate families. People interested in plants come from all over the world to see these primitive curiosities growing in their natural habitats.

The eucalypts and acacias have evolved into major groups unparalleled anywhere else in the world. The eucalypt-dominated forests and woodlands occur in northern and eastern Australia where seasonal rainfall exceeds 200 millimetres in summer, and in southern and south-western Australia where it exceeds 100 millimetres in winter. While eucalypt forests and woodlands are characteristic of the moister coastal and semi-arid

The Fiery Continent

Australia is the driest inhabited continent but, when rains come, plant growth can be rapid and dense. Leaves, twigs and bark accumulate on the woodland floors, and grasses, herbs and shrubs cover the more arid regions. Inevitable droughts subsequently set the stage for another player among the forces shaping our vegetation—fire. Without water, dead plant material does not decay, and creates a fuel load awaiting the next spark.

Bushfires have a history that predates human arrival. Charcoal fragments in lake sediments show that fires, probably triggered by lightning strikes, have occurred spasmodically over the last 250,000 years. However, the frequency and pattern of burning has been most affected by Aboriginal and, more recently, European people. Firestick farming has been practised for at least 40,000 years. It cleared the undergrowth and provided nutrient-rich ash and sunlight for grass and herb growth that, in turn, attracted and supported game.

Vegetation has evolved in response to fire in a variety of different ways. Many eucalypts are encased in thermally insulating bark, beneath which are dormant buds that sprout when the canopy of the tree is destroyed. In other plant types most of the above-ground stems die following hot fires but regrowth is rapid from buried swollen stems called lignotubers. In northern Australia, spinifex grasses will regenerate from the leaf bases while, in eastern parts of the continent, bracken fern resprouts from subterranean stems.

An alternative strategy is seen in those species where the adult plants succumb but the next generation develops rapidly from seeds in the freshly cleared and fertilised soil following the fire. Seeds may avoid destruction, particularly where the soil is low in humus and so does not burn, or where the seeds have been carried underground by foraging ants. In some species, such as many Banksias *and eucalypts, the seeds are retained on the parent plant, enclosed in woody fruits that only disgorge their offspring in response to fire.*

The ability of plants to exist in some sort of equilibrium with fire is highly dependent on the frequency and pattern of burning—the fire regime. If fires recur too frequently, young saplings may be destroyed before they have matured and set their own seed. Eventually, the reserves of seed in the soil are exhausted and the species becomes locally extinct. Alternatively, if fires are too infrequent, seeds may die trapped within their woody fruits as their parents age and fall. There is evidence that European settlement has profoundly altered the fire regime in many parts of the continent. Aboriginal burning was relatively frequent and of low intensity, maintaining open woodlands. Because of the danger to property, stock and human life, Europeans have often attempted to minimise fire, resulting in accumulation of fuel load and less frequent but more severe fires. A greater understanding of the interaction between plants and fire is essential to the management of our forests and grasslands for conservation, recreation and commercial harvesting.

One aspect of Australian vegetation is apparently paradoxical. In a fire-prone environment why are so many shrubs and trees notably combustible? Bark and dead twigs festoon woodland trees, leaves are dry and rich in volatile and flammable waxes and oils. The answer to this paradox may come from the evolution of fire resistance. It benefits those species that can survive fire to promote periodic burning as this eliminates fire-sensitive competitors and opens the understorey for the growth of the next generation.

regions of Australia, *Acacia* shrublands replace eucalypts in dry regions of the continent. These occur in central and western Australia where the seasonal rainfall is lower. The extensive 'mulga' and brigalow scrub of Queensland and New South Wales are familiar examples of *Acacia* communities. There are other, less-widespread vegetation groups: the herblands of northern central Queensland, perennial grasslands, heaths and the shrublands dominated by chenopods, members of the saltbush family of plants.

The flowers of some Australian orchids resemble female insects to lure unsuspecting males into attempted copulation. This orchid mimics the flightless female of a wasp species, not only in form but, more importantly, in scent. The male wasp grasps the female mimic and attempts mating. While such sexual liaisons are unsuccessful for the wasp, the orchid benefits as the male becomes the unwitting vehicle carrying pollen to other flowers.

Another feature of many different types of Australian bush are the grass trees or *Xanthorrhoea* species. These can be found from wet sclerophyll forest through open heath and woodland to the large desert species, *X. thorntonii*. Individuals of these plants can live for up to 900 years and, not surprisingly, they are highly resistant to fire. New leaves rapidly sprout from the top of the blackened stumps and many produce a flowering spike only after fire.

Although the orchids form one of the largest plant families and are found around the world, Australia harbours a great variety of the most interesting and bizarre species. This is because most Australian orchids are firmly rooted in the ground. A very large proportion of the orchids found elsewhere in the world are epiphytes, rooted in the trunks and branches of trees. Another reason is that Australian ground orchids exhibit a unique variety of the strangest possible pollination mechanisms, collectively called sexual deceit mechanisms.

Scientists come to Australia to observe the flowers of ground orchids that apparently resemble female insects. We say 'apparently' because male insects court them and try to copulate with them. In many cases the resemblance between the female insect and the flower is fairly obvious but in others it is very obscure, at least to the human eye. As the male attempts copulation with the flower, pollen is surreptitiously placed on its body. When it has finished, the male will be attracted to another flower where its movements will place the pollen on the correct surface for fertilisation. Fertilisation in the insects only occurs when the male gets it right and finds a real female. These pollination mechanisms mostly involve wasps of various kinds but Australia also boasts the only sexual deceit mechanism that involves ants—male bull-dog ants. In this case, males of a small bull-dog ant species are deceived by the flowers of the Hare Orchid, which release a scent that evidently resembles the scent of the female ant. A male finds the flower very attractive and quickly mounts the lower petal, appearing to gain confidence by its shape and texture. However, its copulatory movements ensure only that a large dab of pollen is placed on its back and within a short time

In many ways the Australian Marsupial Mole resembles the Golden Mole of Africa, although they are not related. The evolution of similar forms in unrelated species is known as convergence.

continues its courtship patrol. Its memory seems to be very short as, even though the next 'female' it encounters may be another flower, it goes through the same routine, its movements placing pollen on a cunningly positioned stigma, ensuring fertilisation of the flower, and picking up a new load of pollen for the next visit.

Some of the most unique ecosystems in Australia are those dominated by spinifex grasses. They belong to two genera, *Triodia* and *Plectrachne,* and, occuping nearly one quarter of the entire continent, grow on poor soils in the arid zone. Each grass plant develops a dome of stems and leaves (for this reason they are also called hummock grasses) but, as the dome ages, the middle dies away leaving a ring. The formation of such domes and rings is very rare outside Australia. They form a spectacular vegetation cover that dominates huge areas, comparable in this respect to the great eucalypt and *Acacia* woodlands. The tough arid-adapted vegetation produced by spinifex grasses accumulates as leaf litter on the ground and provides food for a wide variety of termites. The termites, in turn, are food for a high diversity of lizard species. Thus, although very dry, the arid parts of Australia are not lifeless. They support a unique set of grasses, which, in turn, support a high diversity of termites and lizards.

Australian Radiations

When European settlers first surveyed the Australian bush they frequently attempted to identify the plants and animals with their familiar northern counterparts. For example, they considered kangaroos to be some form of giant mouse and consequently placed them in the genus *Mus,* which includes the European House Mouse! Trees were named oak, beech, pine and ash; birds were called wrens, robins and flycatchers; and fish were named bream and perch.

In most cases, these names could not have been further from the truth. The resemblance of native Australian species to their northern counterparts is a product of an evolutionary process called convergence or parallelism. The principle behind this idea is that each continent possesses a similar array of opportunities and lifestyles for its plants and animals. Species adapted to these habitats may evolve remarkably similar body forms despite great differences in their ancestry.

There is no clearer illustration than the comparison of the evolution of the Australian marsupials with their placental cousins. The plains kangaroos fulfil the role taken by antelopes elsewhere; quolls play the same roles as civet cats; gliding possums parallel the flying squirrels; numbats are anteaters; and the little *Antechinus* are the shrews. The marsupial mole shows a remarkable resemblance to its placental equivalent in Africa, the Golden Mole. The lesson seems to be that, whatever its ancestry, if an animal is going to adopt a mole way of life it will eventually evolve to look like a mole.

Some Australians still have the mistaken impression that our biodiversity migrated here from the Northern Hemisphere. Some did but this view is often incorrect, as is clear in the case of birds. The ancestors of the Emu and cassowaries (and probably the parrots, pigeons and mound-builders) were present when Australia left Antarctica. The ancestors of the passerines, or perching birds, may have originated from the north, in South–East Asia, as long as 50 million years ago (although recently it has been argued that these too may have been Gondwanan). However, the diversity of modern passerines within Australia is almost entirely home-grown. Nearly all Australian passerines are more closely related to each other than to species of similar appearance elsewhere. Australian treecreepers are more closely related to bowerbirds than they are to treecreepers in the Northern Hemisphere. Australian wrens are related to honeyeaters, while Australian robins, whistlers and fantails are related to crows. The relationship of the crow family to the bird fauna of the world is particularly interesting. Recent research suggests that this family (Corvidae) originated within Australia and then spread to the rest of the world.

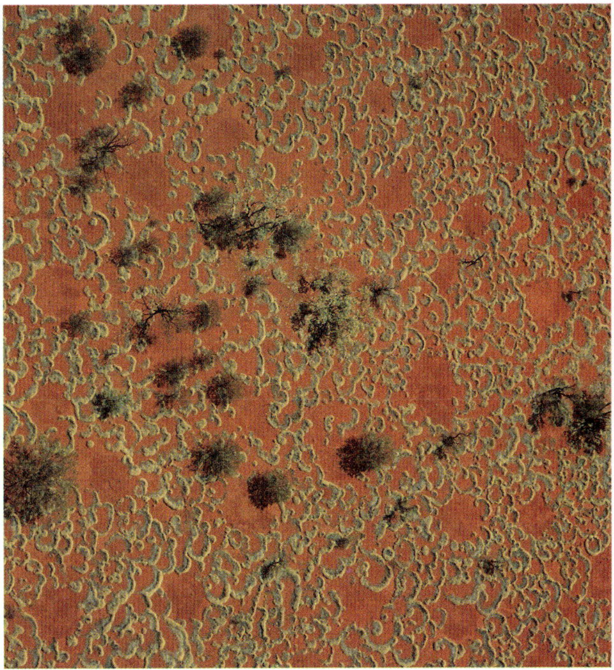

Spinifex ecosystems occupy almost a quarter of the entire continent, on poor soils in the arid zone. They provide food for a variety of termites, which are eaten by numerous lizard species. Thus the arid parts of Australia maintain surprisingly high biodiversity.

Equivalent to the radiation of the passerines is the diversification of those most Australian of all trees, the gums. The familiar blue-green haze of a eucalypt forest appearing in vistas around the continent might look similar from one place to the next but there are actually over 500 species. Eucalypts have gained a foothold in an extraordinary variety of habitats: from the cold windswept hills of Tasmania to the tropics of Cape York, and from lowland swampy coastal plains up to the tree line in the snow-capped Australian Alps. A remarkable feature of eucalypt evolution is that individual species have adapted to surprisingly small geographic areas. Fewer than 10 per cent of species could be described as being in any sense widespread and only the River Red Gum *(Eucalyptus camaldulensis)* has truly colonised the entire mainland.

Suggested Reading

Barker, W.R. and Greenslade, P.J.M., 1982.
Evolution of the flora and fauna of arid Australia.
Peacock: Frewville, South Australia.

Gill, A.M., Groves, R.H., and Noble, I.R.,1981.
Fire and the Australian biota.
Australian Academy of Science: Canberra.

Groves, R.H., 1981.
Australian vegetation.
Cambridge University Press: Melbourne.

Groves, R.H., and Burdon, J.J., 1986.
Ecology of biological invasions: an Australian perspective.
Australian Academy of Science: Canberra.

Keast, A., 1981.
Ecological biogeography of Australia.
Junk Publishers: The Hague.

Sabath, M.D. and Quinnell, S., 1981.
Ecosystems energy and materials: the Australian context.
Longman Cheshire: Melbourne.

Frontiers in Biodiversity

Biodiversity is a new science that faces many exciting challenges, not the least of which is determining how much diversity there is. While the study of biodiversity has its roots in ecology and evolution, it has recently emerged as a distinct science with its own special set of questions. In many cases the methods required to answer them are still being developed and many answers will be discovered only by painstaking research. This chapter explores some of the exciting challenges that face the science of biodiversity.

How Many Species are there on Earth?

A frequent answer to this question is 'I have not got a clue'. If that was your response, you are in good company because nobody knows and scientists can only make very rough estimates. It is an interesting commentary on modern science that we are intent on exploring the universe but know so

ABOVE *The minute size of many species brings problems of sampling and technology. Here a biologist inspects leaf litter on a forest floor for invertebrates. Very little can be seen with the naked eye but research has shown that beneath his feet there may be as many as 100 species of invertebrates and a vast array of micro-organisms.* **LEFT** *Beetles make up an extraordinarily diverse group of animals. In one South American study, 1,200 species were found in the canopies of just 19 trees. A similar diversity of Australian beetles may exist but this remains unknown until further research is carried out. This Fiddler Beetle is feeding on a eucalypt blossom in a Sydney backyard.*

SPECIES DISCOVERIES

Vertebrates are much better known than invertebrates. These curves represent the rates at which new species are being discovered. The vertebrate curve (green) flattens out showing that, while new species are still being found, this occurs fairly infrequently. By contrast, the invertebrate curve (yellow) is getting steeper as new species are being discovered at an increasing rate.

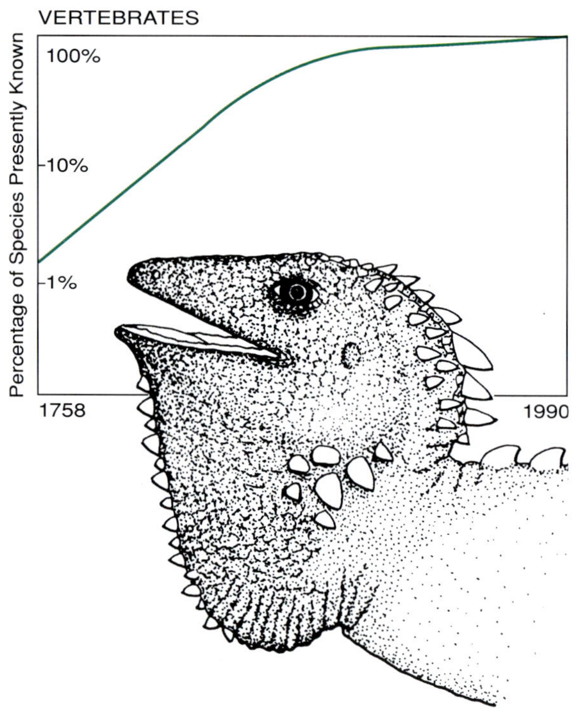

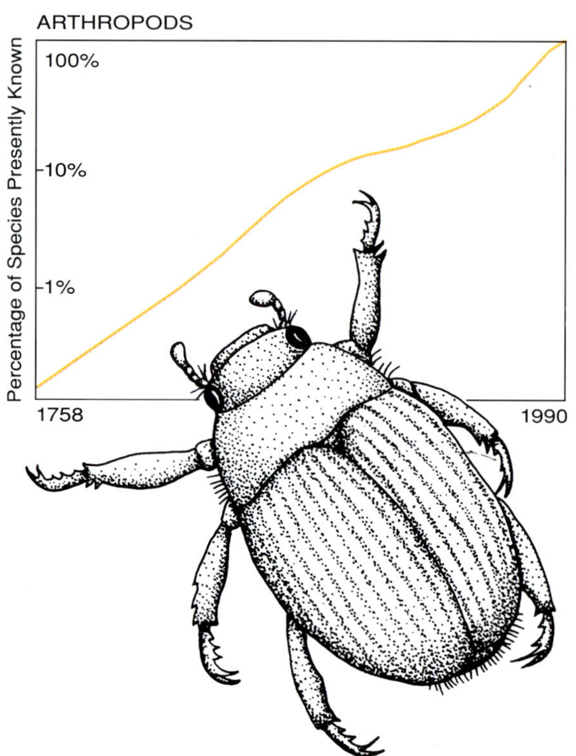

little about the number of species with which we share our own planet.

The challenge of counting all the species in the world is complicated by the fact that the process of discovery is still going on. This means that at the moment there are four categories of species:

• Species for which specimens have been collected from the wild and formally named and described in the scientific literature. There are currently about 1.82 million in this category. This figure is not exact because, even though we have central international catalogues for all the books in the world, a single central catalogue for species does not yet exist.

• Species that are lodged in research collections but have not yet been examined carefully enough to be formally named and described.

• Species thought to exist in the wild but specimens have not yet been collected. Most of the current debate about the number of species in the world concerns the size of this category.

• Cryptic species: nearly all classification is based on physical features apparent in specimens stored in collections. Sometimes there are suspicions that what was believed to be a single species may actually be a group of species that closely resemble each other. Further detailed studies, often involving the biochemical analysis of specimens, may reveal that this is indeed the case. Such groups are known as cryptic, or hidden species. The Australian velvet worms (Peripatus) are a good example. Biochemical analysis has shown there are over 90 species and not just six as previously supposed.

Previous Estimates

Until the 1980s there was general agreement that there were three to five million species in the world. This figure was arrived at by the following steps:

• The insects were thought to be the group with the most species and about one million insects had

Do Tiny Differences Between Species Matter? The Water Fern and the Weevil

Salvinia *is an introduced fern that floats on the surface of water. In many warm parts of the world, including Australia, it is a major pest. It clogs lakes, rivers, dams and canals, preventing travel by boat and cutting off the sunlight to the water below, disrupting aquatic ecosystems and killing fish.*

ABOVE AND BELOW Salvinia molesta *infesting waterway: before and after biological control.*

Biological control involves finding a herbivore that normally eats the pest plant in its native country. This herbivore is then introduced to the pest-affected area. An infestation in Queensland was at first thought to be the species Salvinia auriculata *from the Caribbean. A long-nosed weevil,* Cyrtobagous singularis, *was brought in from the Caribbean to Queensland but it had little effect on the water fern.*

Further investigation led to the discovery that the pest species was in fact Salvinia molesta, *which looked almost exactly like* Salvinia auriculata *but came from Brazil. Although the differences in the appearance of the two water fern species were minute, they resulted in relocating the search for a herbivore from the Caribbean to Brazil. In Brazil, another weevil,* Cyrtobagous salviniae, *which was almost identical to* Cyrtobagous singularis, *was found eating the water fern. It was introduced to Queensland where, after appropriate quarantine trials, it quickly devastated the water fern. In a single year, its population grew from a few thousand to over 100 million and destroyed 30,000 tonnes of* Salvinia.

The water fern example shows that closely related species are biologically different in important ways. In this case the importance was economic as well as scientific. Put another way: it is definitely not true that if you have seen one weevil you have seen them all! It also shows that the all-important differences in the fern and weevil species would have been missed without the existence of biologists who understand fern and weevil diversity.

Estimating Host-specificity Among Beetles on *Luehea* Trees

New estimates of the number of species in the world began with Erwin's data on the beetles found on just 19 Luehea seemani *trees. His collection contained an astonishing 1,143 species, a number he eventually rounded up to 1,200 because he had missed some of the weevils. Then, based on experience from previous research, he estimated that 20 per cent of the herbivorous beetles were specific to this host tree. Thus, the first line of numbers shows that 20 per cent of 682 herbivorous species—that is 136.4 species— were host-specific. Similar reasoning was then applied to the predatory, fungus-eating and scavenging beetles to produce a grand total of 163 species specific to this one tree species.*

Type of beetle on Luehea	*Number of species found*	*Host specificity %*	*No.*
Herbivores	682	20	136.4
Predators	296	5	14.8
Fungus-eaters	69	10	6.9
Scavengers	96	5	4.8
Totals	1,143 (1,200)		162.9 (163)

been named and described, most of them from the better-known temperate-zone environments of Europe and Northern America. However, it was clear that there were plenty of other insects not yet collected or described, especially in the tropics.

- Most bird and mammal species, both in the temperate zones and the tropics, were already known. In these groups there were two to three times as many species in the tropics as in temperate zones.
- When this tropical–temperate ratio derived from birds and mammals was applied to the insects, it led to the estimate of two to three million insect species in the tropics, most yet to be described.
- By adding to this the one million insect species from temperate zones, plus the contributions from all other plant and animal groups, an estimated total of three to five million species worldwide emerged.

Recent Developments

Within the last few years, the estimate of three to five million species has been challenged by the entomologist Terry Erwin and his colleagues working in the tropical rainforests of South America. Using insecticide he fogged the canopies of just 19 specimens of the medium-sized evergreen tree, *Luehea seemanni,* and caught all the insects that fell to the forest floor. The results were breathtaking and plunged the scientists into a new round of frantic research.

A very large proportion of the insects found had not been collected or described previously. There were so many unknown that the research team concerned themselves with just the beetles. There were 7,735 specimens that were classified into 1,200 species.

These staggering numbers prompted new estimates of the total number of species in the world. The critical question was: how many of these beetles were host specific, that is found on *L. seemanni* trees and nowhere else? The reason for this concern was that if each tree species in the forest has its own unique set of beetles, then the number of beetle species alone could be immense.

Erwin came up with a total of 163 host-specific beetles (see table above). This started a chain of reasoning to calculate the total species in the world as follows:

- There are about 50,000 tree species in the tropics. If each of these harbours, on average, 163 specialist beetles, there will be a total of 8,150,000 specialist beetle species.

• Erwin made the assumption, based on earlier research, that beetles make up 40 per cent of all arthropod species. Using this figure, the total number of arthropod species came to 20,375,000.

• As this estimate is based on sampling only the tree canopy, we must account for the ground beneath it. If we assume that canopy arthropods have twice as many species as ground arthropods, we then add another 10 million species. This provides a grand total of 30,375,000. We can round off this estimate to 30 million.

Notice that the original estimate for all species was three to five million but now, working with recent data, there is a ten-fold increase *for the arthropods alone.*

At this point it is very important to inject a note of caution. If the assumptions Erwin made were wrong, then the total would change. Several entomologists have challenged his assumptions, especially on the rates of host-specificity and agree that the total number of arthropods is more likely to be 10 million. This may or may not be correct but it is still twice the previous estimate for all organisms put together. Clearly there is a need for basic research into his assumptions.

One approach to estimating the numbers of some of the other organisms is to consider the parasites. If we assume there is just one species of parasite (a protozoan, roundworm, fungus or bacterium) that specialises on each of the arthropod species then the estimate of the total number of species immediately doubles. In fact, species that have been thoroughly studied usually have more than one specialist parasite. If there are five parasite species for every arthropod species, which would not be unreasonable, the estimate for the total increases fivefold.

If we then add the other organisms—such as non-tree plants, the other (non-parasitic) bacteria and fungi and all the other invertebrates—we add several million more.

Which Groups are Still Largely Unknown?

Free-living Bacteria

Several thousand species have been named but the usual methods of describing organisms cannot apply to bacteria as they have no obvious features like the wing-veins of insects or the colours of birds.

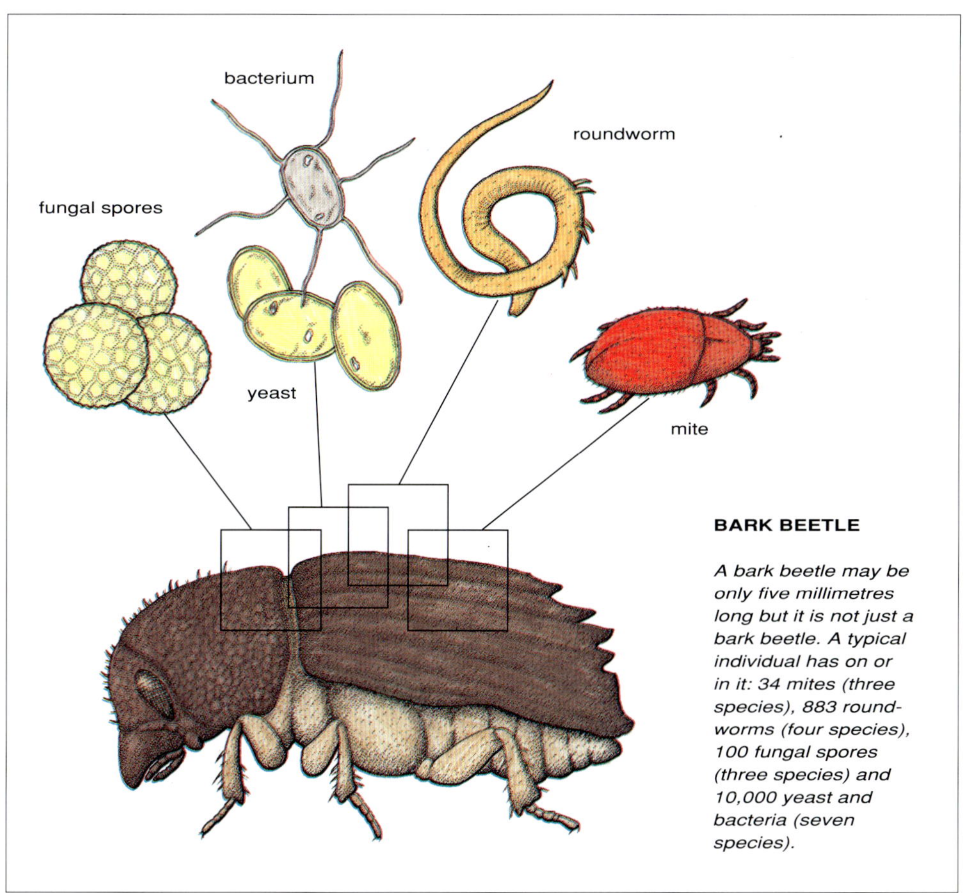

BARK BEETLE

A bark beetle may be only five millimetres long but it is not just a bark beetle. A typical individual has on or in it: 34 mites (three species), 883 roundworms (four species), 100 fungal spores (three species) and 10,000 yeast and bacteria (seven species).

Since 1989 new molecular methods have made it possible to find bacterial DNA in the environment and then read the genetic information it contains. Two studies published in 1990 illustrate the radical results that are emerging. Both studies were of environments where a lot of effort had previously been made to culture and identify bacteria: one in the hot springs of Yellowstone National Park, USA, and the other from the Sargasso Sea (a part of the Atlantic Ocean). DNA was obtained from water samples and 17 distinct bacteria were found. The startling result was that none of them were the same as anything that had previously been cultured from these environments or indeed from anywhere else. This means there could be any number of bacterial species not discovered by culturing. In the Sargasso Sea study, several of the bacteria found were genetically very distinct from any others previously known. The genetic difference was as dramatic as that between, say, insects and worms.

Colourised scanning electron micrograph of Escherichia coli. *This micro-organism is undoubtedly the most intensively studied because it lives within the human intestine, yet there is still much to learn about it.*

Protozoa

A few thousand protozoa have been described to date but it is widely accepted that this is only a small proportion of the total. Although estimates vary between 65,000 and 200,000 species, it has been suggested that many animals and plants have at least one protozoan associate. That would imply many more species.

Nematoda

About 80,000 species of nematodes have been described but taxonomists currently estimate that there may be a million. They live in many different places, some free-living in soil and water, others parasitising plant roots or the skin, lungs and stomachs of animal hosts. The reproductive capacity of some species is legendary with females laying 200,000 eggs a day.

Mites

To date about 30,000 species of mites have been described but there may be millions more to go. They occupy the most bizarre and unexpected habitats, including the breathing chambers of snails, the noses of mammals and the wing-cases of beetles. Astonishingly, one species lives in the ears of moths but this mite only ever infests one ear so that the moth does not go deaf and can continue to detect its enemies!

Most of these groups have two characteristics: first, they use other organisms as their habitat, often specialising, making a particular species their home. Thus, if there are many host species available, for example, 10 million insects, then the

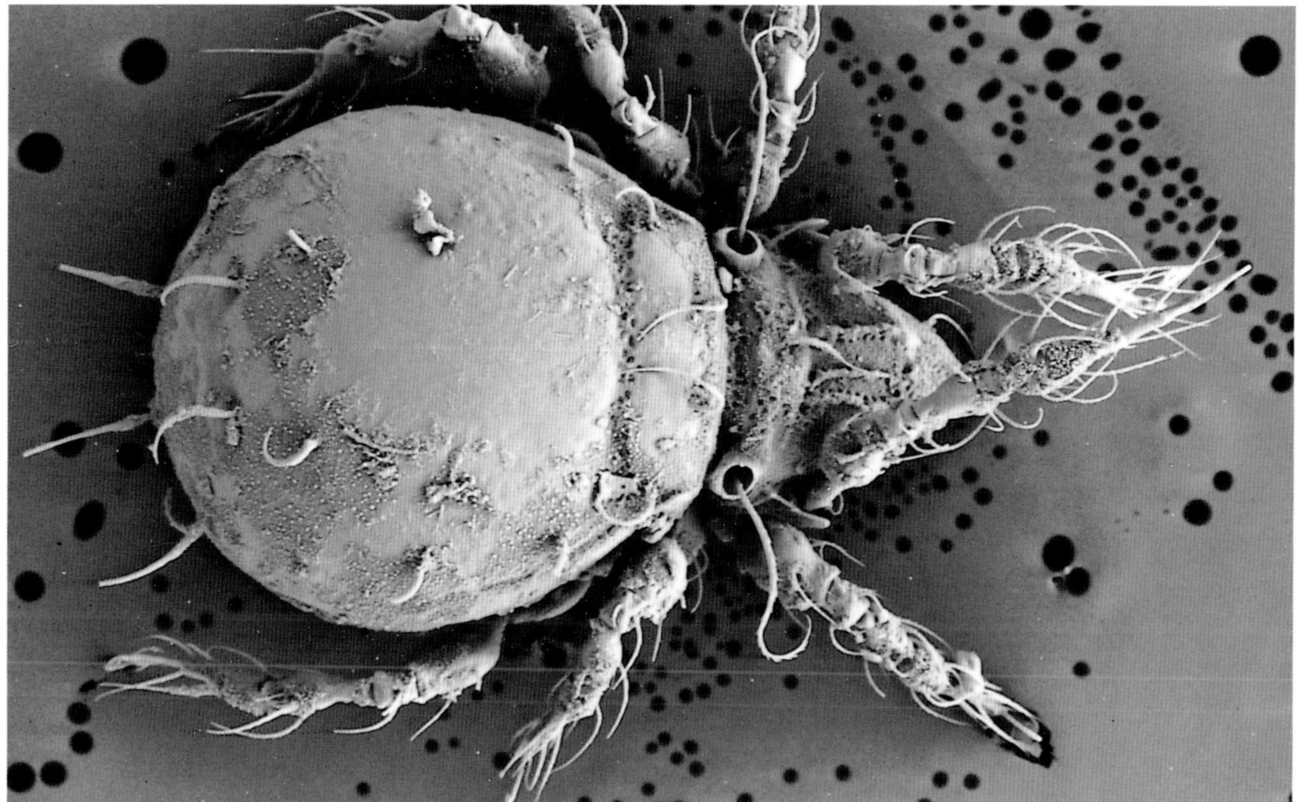

evolution of an equal number of bacterial, protozoan, nematode and mite species is possible. Second, unlike the insects, they often have relatively few obvious features that can help us distinguish one species from another. Thus, it is possible that significant variation lies hidden among sets of organisms that look very similar, even under the microscope.

Where are Most New Species Being Found?

New species are being discovered every day in many different kinds of habitats. At present it appears that there are two sorts of environments where very large numbers of new species may be expected: tropical rainforests and the oceans. Each of these environments contains many different habitats and exploration of them is still in its infancy. Many other environments are thought to be relatively well known but even the most familiar places are worth another look using new technology.

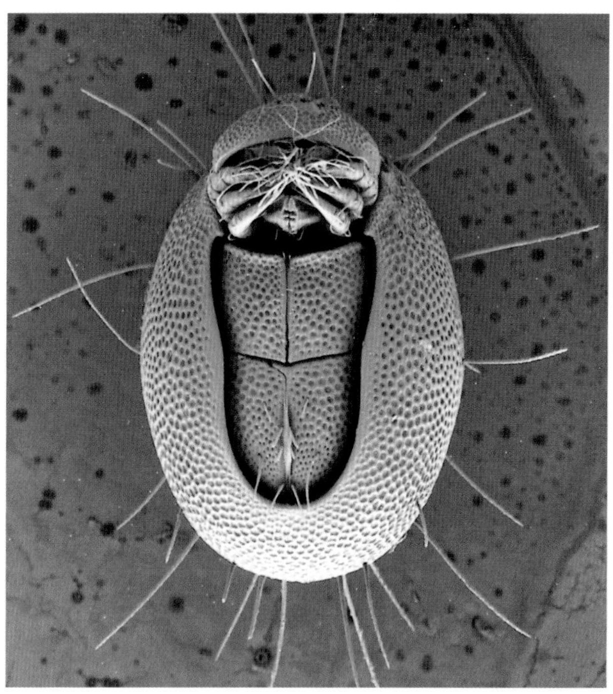

Although mites are generally tiny, barely visible with the naked eye, they exhibit an astonishing diversity in both form and habitat. Some, like these, are important soil decomposers. Others consume fungi and bacteria that grow on leaves. Even the best-groomed people harbour mites in their hair follicles and we share our homes with dust mites.

Biodiversity is particularly high in coral reefs. Although we are rapidly increasing our knowledge of coastal marine biodiversity, much remains unknown.

The Oceans

Almost three-quarters of the surface of the Earth is covered by ocean, much of it very deep. Compared with land environments, ocean biodiversity beyond the continental shelf and coral reefs remains relatively unknown. It used to be thought that the deep ocean floor and the open ocean supported few organisms. However, recent samples taken from the ocean floor in Bass Strait and off the coast of New Jersey have shown that hundreds of species live there and that what were once regarded as ocean deserts are, in fact, teeming with life (see Chapter 6).

In the open ocean, new techniques have detected a vast new array of organisms, all about one millionth of a metre in size, each consisting of a single cell. These organisms are known as picoplankton or ultraplankton. In many parts of the world, what appears to be clear water is, in fact, a soup of these tiny organisms. As yet, it is not known what they eat or what part they play in the ocean ecosystem. In addition, very recent studies suggest that viruses may also play an important part in marine food webs.

Volcanic activity on the seabed often produces areas of warm water known as hydrothermal vents where organisms can live, despite great depths that are devoid of sunlight. In the last few years 190 new species of marine animals have been described from these habitats (see box on page 58).

Forest Canopies

Sampling species from the rainforests of equatorial South America and Queensland has shown that the tree canopy high above our heads is another world, harbouring hundreds of thousands of unknown species, most of them insects. The number of new species among other arthropods, invertebrates and micro-organisms in the forest canopy remains unknown.

Why is Biodiversity so Hard to Find?

There are two major challenges to discovering our total biodiversity. First, most biodiversity consists of very tiny organisms. Second, many organisms live in habitats that are tricky to get at—some habitats, like tiny spaces between grains of sand, are so unexpected they are difficult to anticipate. A third problem is that some organisms such as bacteria are difficult to separate into species.

In general, the vertebrates (mammals, birds, reptiles, amphibia and fish) and the flowering plants are catalogued and counted quite easily. However, even this can be a big task: for example, a research team counted 330 species of flowering plants and ferns in one small backyard in Sydney. The majority of species—the invertebrates, lower plants, fungi and bacteria—are often exceedingly difficult or impossible to identify. Here we will take a brief look at the first two problems; the third is described in Chapter 5.

The problem of minute size is also a problem of sampling, technology and, consequently, funding. Imagine yourself leading a biodiversity survey team. Suppose you take 100 samples of soil and leaf litter from the floor of a tropical forest—you catalogue all the invertebrates in them and you are still finding many new species. Suppose this took a year, a team of six full-time scientists and you have found 10,000 species, 6,000 of which are new to science. Do you stop? If you decide to catalogue the fungi, you now require a large microbiology laboratory with its culture techniques and electron microscopes. After another year your team of microbiologists has identified 1,000 fungal species and every time they add another sample at least 25 more are detected. Do you stop? Your two-year budget is already over half a million dollars excluding the buildings, equipment and instrumentation.

Suppose you do stop there. A major question remains: if you have 11,000 species of invertebrates and fungi at one place, how different are they at another? To find this out, the entire exercise must be repeated at an adjacent site. This is an extremely important question because the rate at which species change across habitats or landscapes makes a major contribution to the biodiversity of

Habitats Worth a Second Look

A team of biologists recently sampled the soil of a forest in the western United States where there had been previous surveys. New technology for examining the material at very high magnifications revealed an array of minute organisms never previously suspected. The surprise was that they were not all bacteria and fungi but many were tiny arthropods—so many, in fact, that this area harbouring just 143 species of reptiles, birds and mammals probably had as many as 8,000 arthropod species.

In another example, freshwater biologists in Australia took another look at life in our ponds, creeks and rivers. A life form new to Australia—one only barely known to science—was discovered in the creek running through the middle of the Macquarie University campus in Sydney. The animal was a millipede, but aquatic rather than terrestrial, with its nearest relatives known only from the Amazon. Other research has shown how little we know of minute Australian freshwater animals. It is estimated that there are over 50,000 species of single-celled freshwater animals (protozoa) in the Northern Hemisphere, but research suggests that the diversity in Australia is even greater.

As technology improves, the biodiversity of many environments could prove much higher than previously thought.

Rapid Biodiversity Assessment

The scientific process of collecting, describing, naming and managing large collections of specimens is painstaking and time-consuming. Frequently, decisions concerning the conservation value of a particular place must be made quickly, often without enough time or resources. Responding to this situation, Australian scientists are developing ways to catalogue the variety of species by simple visual inspection, either by eye or using a simple microscope. The 'species' identified by this method are called Recognisable Taxonomic Units or RTUs and each one is given a number. As new specimens are added, they are either assigned a pre-existing number or, if they are new to the collection, given a new one. In many situations it does not matter whether or not a specimen has a scientific name for biodiversity surveys to be successful: a number will do. So far tests have shown that technicians with only a half-day of training can quickly become expert at sorting hundreds of specimens into RTUs that correspond quite closely to the species found by professional taxonomists.

the area (see box on page 74). If you obtain one set of species from one site but another set from a spot one kilometre away, then your biodiversity is immediately doubled. Trials of this kind have shown that in some kinds of environments you have to move only a few kilometres to find an almost completely new set of species. Since such environments pack so much biodiversity into the space they occupy, they may be important to conservation. But, for the same reason, they are the most expensive to discover in the first place.

The second problem concerns the habitats of many species. One way of understanding this problem is to think about parasites. Many people encounter parasites only occasionally—a tapeworm in their dog, ticks from a bushwalk or head lice among school children. In fact, parasites are among the most common and diverse of all organisms inhabiting the nests, exteriors and internal organs of all other organisms. If there are 10,000 invertebrate species on our list, many will harbour at least one host-specific parasite thereby doubling the biodiversity. In fact, many organisms harbour at least one parasite from each of the following groups: bacteria, fungi, roundworms, single-celled animals (protozoa) and arthropods. This means that in many cases what appears to be one species is in fact a tiny community made up of six or more species.

At the start of this section we noted that many

habitats are difficult to anticipate. How difficult? One small group of mite species lives exclusively under the wing-cases of certain beetles. A small group of wasp species lives exclusively in figs. Many single-celled organisms are restricted to certain parts of the intestine of some termites. Thus, beetle wing-cases, figs and termite intestines are habitats. Any part of any animal or plant may be home to some species—the brain, kidney, gill, hoof, flower, leaf, root or trunk, to name just a few. With so many possibilities, it is difficult to know where to stop looking. Finally, we have very little idea how bacteria and fungi divide the world into habitats. These mostly microscopic organisms present perhaps the greatest challenge of all.

By now it should be clear that the cataloguing of biodiversity, even just the immediately useful biodiversity, is a project as vast and expensive as those encountered in astronomy and particle physics. They are alike in many ways, not least is the fact that it involves objects that are not easily seen and therefore not easily imagined. Cosmic bodies, subatomic particles and micro-biodiversity all require enormous resources to explore and understand. Most developed countries are prepared to build particle accelerators and radio telescopes. Where are the biodiversity laboratories?

Suggested Reading

May, R.M., 1991.
A fondness for fungi.
Nature 352: 475–476.

May, R.M., 1992.
How many species inhabit the Earth?
Scientific American 267: 18–25.

Moriarty, D.J.W. and Veal, D.A., 1992.
Increasing awareness of microbial ecology.
Search 23: 100–103.

Olsen, G.J., 1990.
Variation among the masses.
Nature 345: 20.

DIVERSITY IN PROPORTION

Diversity of different major groups as represented by the size of the organism. Represented here by a kangaroo, frog, lizard, cockatoo and fish, vertebrates show relatively little diversity compared to worms (shown by the earthworm), single-celled animals (shown as a footprint-shaped Paramecium*), fungi (toadstool) and the molluscs (snail). Plants are also diverse but still dwarfed by the arthropods, shown here as a lobster and a bull-ant. It is not yet clear how the bacteria should be represented. It may be that they, in turn, dwarf the arthropods.*

Managing Biodiversity

The processes of speciation and extinction are usually slow and it takes millions of years for new species to replace those that were lost. The most significant problem for biodiversity today is habitat destruction, leading to the accelerated rate at which humanity is causing extinctions. Species are being lost without being replaced. To secure the future, an understanding of human impacts on biodiversity is required together with ways of eliminating or minimising them.

Recent Impacts

Australian ecosystems have been changing ever since life first colonised the continent. Following the arrival of humans, the rate of change has been both accelerated and diversified. In recent decades, the complete or partial removal of vegetation for pastures, crops, forestry, urban development and mining has been widespread. Such changes have contributed to the loss of soils, habitats and species. Soils have also been acidified by fertilisers, salinated by irrigation, poisoned by pollutants and eroded by land clearing and overgrazing. There have been major changes in wildfire patterns that have altered entire ecosystems as well as the distribution of many individual plant and animal species. Added to this scenario, hundreds of foreign species have been introduced into Australia, many of which have become widespread at the expense of their native counterparts.

The speed at which many changes have occurred caused Don Adamson and Marilyn Fox, two Australian ecologists, to write: *'Virtually the whole continent is in flux because the new impact is open ended and because not all changes instituted*

Many vertebrates are vulnerable to extinction. The tiny Corroboree Frog is confined to the coldest and moistest part of the mainland, in areas above 1,000 metres in the Snowy Mountains. The impact of climate change may gradually eliminate its habitat. A puzzle exists over worldwide declines in numbers of frogs—some species appear to have vanished very recently.

earlier have worked through the system. It is salutary to realise that the life-span of individuals of many Australian plants is greater than the total time since the European arrival, so that some plant communities contain abnormal numbers of ageing individuals inherited from before European settlement. In such vegetation, hidden changes have yet to become fully manifest.'

Various human activities have wiped out 20 species of small mammals in Australia. Others, like the Bilby, are now highly endangered.

Patterns of Extinction in Australia

Vertebrates

Twenty species of mammals and 10 species of birds have become extinct since 1788. Most of these extinctions have occurred in the drier regions of the continent. However, it is difficult to attribute the demise of any species to a single cause as different vertebrates are affected by different combinations of factors. Take, for example, the decline and extinction of medium-sized mammals following European settlement. Many species of ground-living mammals from about the size of a large mouse up to that of a cat (35–5,000 grams) have recently become extinct in arid areas of Australia. This strangely selective set of extinctions has been difficult to explain. However, a recent theory goes as follows:

Most mammals in this size class are plant-eaters and arid Australia is a rather hostile environment because of its poor soils and sparse vegetation. Much plant food is confined to relatively small fertile areas, which, during periods of drought, become crucial refuges for these native mammals. Following the introduction of rabbits and cattle, which quickly became extremely numerous, this situation broke down as, during droughts, they consumed most of the food in the refuges. Native mammals found themselves in the minority and were pushed further towards starvation. Introduced predators like foxes and cats exacerbated the situation by preying on native and exotic animals alike. These factors together eventually wiped out 20 species of small mammals. Others, such as the Bilby, are presently endangered.

The habitats of ground-living mammals and birds have been greatly affected by scrub clearance, trampling by domestic stock, overgrazing and soil erosion. Historical records show that many local extinctions followed waves of settlement across Australia. More extinctions followed the introduction of foxes, rabbits and the descendants of domesticated stock and pets, including pigs, goats, horses, camels, dogs, cats and buffalo, which had either escaped or been abandoned. A variety of other vertebrates are considered vulnerable to extinction, including species of fish, frogs, turtles, lizards, skinks and snakes.

Invertebrates

As the benefits of invertebrates are only just being widely recognised, there have been few attempts to record their fate in the past. A number of invertebrates, including some dragonflies, butterflies and large beetles, are known to be in serious decline. The threats to their existence vary from the pollution of rivers and lakes where larvae live, through forest clearance and logging to urbanisation, which may simply cover crucial habitats with concrete. Many freshwater crayfish are known only from single streams or particular stretches of rivers and are considered to be very vulnerable.

The removal of vegetation, even when it is replaced by crops or plantations, almost certainly removes most of the invertebrates that are dependent on it. As a very high proportion of Australian plant species are found nowhere else, the invertebrates that are highly specialised for living on them may be especially vulnerable. The introduction of domestic stock causes compaction of soils and changes in soil fertility. One study of the important soil invertebrates known as springtails showed that pasture improvement and grazing encourages invasions by species from overseas, which then grow like animal 'weeds', displacing native springtails.

Plants

The history of plant diversity in Australia is better documented. Ninety seven plant species have become extinct within the last 200 years. How many of these extinctions were directly the result of human activities is open to conjecture, but considering there are a further 3,329 rare or threatened plant species (about 17 per cent of the flora) today provides strong evidence that current land-use practices place very significant numbers of plant species in jeopardy.

Some areas are more affected than others. Forty-three per cent of all our rare and endangered plant species are in Western Australia, the majority in the south-western corner of the State. Cape York, the Moreton district of south-eastern Queensland and the southern tablelands of New South Wales all have particularly high numbers of plant species considered to be vulnerable to extinction. They include trees and shrubs, orchids, grasses and lilies.

A word of caution may be necessary here. Many plant species are naturally rare, for example, being restricted to a particular type of soil. Some of these have probably always had very local distributions. This means that the situation is not simple: some rare plant species are not in any danger, while some that were once very common, such as many mallee and native grassland species, are in imminent danger of extinction.

Climate Change and the Greenhouse Effect

An unknown chapter in the future of biodiversity is that concerning climate change. This may have a far greater effect on biodiversity than any other process. In general, if the entire Earth becomes slightly warmer, species that are sensitive to temperature changes will tend to migrate to find more suitable habitats. Therefore, as the climate changes, forests and farmlands currently managed for production will be crucial as reservoirs of biodiversity. This is not only because they cover much larger areas than specifically designated conservation areas but also because species will migrate to them. Landowners of all kinds will, therefore, play a pivotal role in the preservation of biodiversity.

What Does the Enhanced Greenhouse Effect Mean for Biodiversity?

Future climate change resulting from the enhanced greenhouse effect may have substantial but, as yet,

What is the Greenhouse Effect?

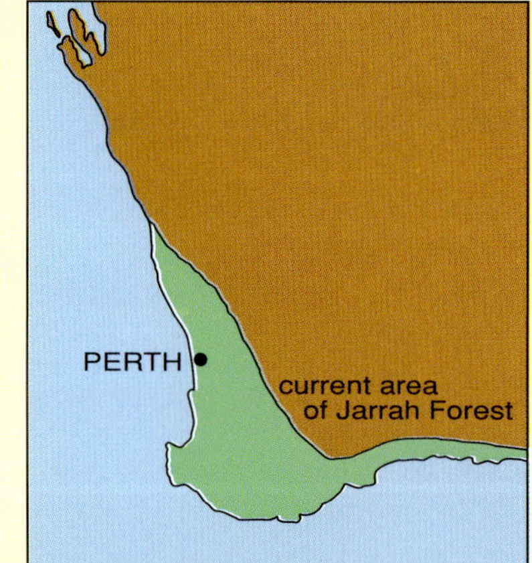

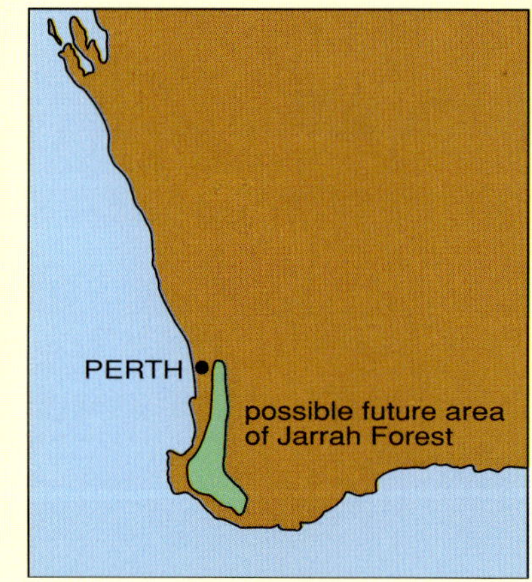

Current and projected limits to the Jarrah (Eucalyptus marginata) *forest of Western Australia. The distribution of this special forest type will be restricted if the climate warms a further three degrees. The areas it now occupies are expected to become a woodland dominated by* Eucalyptus wandoo, *which may benefit the Numbat but be deleterious for a variety of Jarrah forest animals.*

Sunlight—short-wave solar radiation—penetrates the atmosphere and is absorbed by the surface of the Earth. The Earth warms up and reflects long-wave radiation back into the atmosphere. This radiation is partially absorbed by some of the gases in the atmosphere, which acts like a nursery greenhouse, warming up and then radiating heat back down to Earth. By returning this heat back to the surface, these 'greenhouse' gases make the atmosphere and surface of the Earth roughly 33°C warmer than they would otherwise be, making it a hospitable place for life to flourish. The main natural greenhouse gases are water vapour, carbon dioxide, methane, nitrous oxide and ozone.

Human activities associated with increasing industrialisation have resulted in increased emissions of greenhouse gases, such as carbon dioxide and methane, and the additional release of synthetic gases such as the chlorofluorocarbons (CFCs). These extra emissions are capable of enhancing the greenhouse effect, causing further warming of the surface of the Earth. If current rates of emission continue, global mean temperature is predicted to increase by about 2.5°C above the present value by 2030 and a possible 5°C before the end of the next century.

unpredictable, effects on biodiversity. Global warming may be accompanied by changes in other climatic variables, such as the amount and timing of rainfall and the frequency of tropical cyclones. Changes in snow and sea levels may also have important consequences. Both the amount and effects of global warming are expected to be greatest in the temperate and arctic regions and least in the tropics.

In Australia, the most recent climate change scenarios produced by CSIRO indicate that average temperature increases of up to 2.5°C may occur by 2030. By 2070, a warming of up to 5°C may occur in some parts of the country. In the northern part of Australia, the temperature rises may be accompanied by an increase in rainfall of up to 20 per cent, while in the south, rainfall may decrease by a similar percentage.

Some species will benefit from the predicted changes, while others will be disadvantaged. Some may be directly affected by changes in factors such as temperature, rainfall and carbon dioxide concentration. Climate change may affect others indirectly by altering habitat, food availability, fire frequency, predator–prey relationships and pathogen attack.

In the past, species have responded to climate

Habitats of ground-living mammals and birds are disturbed by scrub clearance, trampling by domestic stock, overgrazing and soil erosion. Land degradation due to overstocking of sheep can be clearly seen on one side of this fence bordering a farm in Western Australia. Natural vegetation exists on the other side.

change in different ways—by shifting to keep track of their preferred climate, by staying in the same place and adapting to a new climate, or by going extinct. The rate at which the climate is predicted to change in the future is much faster than any change recorded in the past. If the predicted changes occur, species will need to migrate 30–100 kilometres per decade just to remain within their current climate. Some, like birds and flying insects, may be able to migrate at this rate but many others will not. Even the more mobile organisms will face the problem that their natural habitat has become so fragmented that they will be confronted with areas of inhospitable habitat that will act as barriers to migration.

Many biological attributes of species will be important in determining how they will respond to climate change. Species that are widely distributed, have short generation times, or are capable of wide dispersal will have a head start. The most vulnerable will be those in small populations, those that have low genetic variability or a very specialised way of life and those that are poor dispersers or are already isolated in small reserves. Because species will respond to climate change at different rates and in different directions, present day communities may be reshuffled and new assemblages may form within a relatively short time.

Certain ecosystems can also be identified that may be particularly vulnerable to future climate change. Alpine habitats, for example, often consist of mountain 'islands' in a sea of inhospitable habitat. To keep up with a 3°C warming, alpine species will have to shift about 500 metres uphill, so many species will simply run out of mountain. Coral reefs and wetlands may be severely affected if the predicted sea level rises of between 10 and 100 centimetres occur during the next century.

Climate change in the future may reduce the ability of reserves to protect the range of species currently within their boundaries, and other conservation procedures will be required within and beyond current reserves. Adaptation to climate change will involve measures such as the provision of corridors and buffer zones to link reserves and the design of reserves with wider geographic and climatic variation. The conservation value of forests and farmlands currently managed for production will become increasingly more important.

Where Should Biodiversity be Conserved?

This is a major question puzzling scientists in Australia and around the world. We have seen that the value of protected areas will change because of accelerating changes in the climate. On top of this, vast areas of Australia's land and coastal seas must remain productive, to feed us and to generate exports. It appears that this dilemma cannot be completely overcome with systems of national parks and reserves. This is precisely what scientists around the world have concluded and there is now general agreement that what is required is bioregional management.

Bioregional Management

A bioregion is an area defined for the management of biodiversity by geography and ecology rather than political or economic criteria. An example of this would be a river drainage system from its mountain springs to its mouth.

Biodiversity is so crucial to human survival that it should be maintained wherever it occurs. This does not mean that the entire world should be converted into a patchwork of reserves. However, it does require that human activities of all kinds—farming, forestry, fishing, industry and urbanisation—are harmonised with this objective. This is the function of bioregional management.

Imagine a valley surrounded by hills, with a broad river that winds down to a mangrove swamp on the coast. When resource management is confined to local issues and not the bioregion, the following hypothetical situations might arise:

Bioregional management is based on geographical and ecological guidelines, as well as on political and economic criteria. The entire region from the mountains to the coast includes many types of land use and several local governments. Under bioregional management, a single authority made up of representatives from all key agencies would be responsible for sustainable production and management of biodiversity.

• Loggers take trees from the hills and soil erosion adds large quantities of mud and debris to the river, making the water unpalatable for livestock kept by the farmers in the valley. The farmers and loggers are at odds.

• A developer obtains permission to fill in the mangrove swamp and cut down the mangroves to build a seaside resort. The fish populations that once spawned there go into a massive decline. The developer and the fishermen are at odds.

• The farmers clear bush for pasture, the developer extends his resort, the loggers replace the native forest with pine plantations and suburbs spread out from the town. People notice that their local biodiversity is in very bad shape. Their gardens are no longer visited by many kinds of parrots, they see fewer native animals as they travel through their valley and, to top things off, locals and tourists who go fishing come home empty-handed.

Bioregional planning takes into account the entire valley and its river drainage, including the coast. In this case the bioregion could be many thousands of hectares, possibly crossing several local government and even a State boundary. It includes production areas for forestry, farming and fisheries and one or more towns and villages. Bioregional planning would provide alternative scenarios such as:

• In the hills, the entire forest is assessed for those areas suitable for plantations, biodiversity reserves and timber extraction. Plantations are created in areas of negligible conservation value. Logging procedures now leave a buffer zone along the banks of the river and very steep slopes are left uncut. These measures reduce soil erosion and river pollution. Loggers leave canopy and nesting trees that ensure the survival of healthy populations of birds and mammals dependent on old growth. A bio-diversity reserve is established to protect the rarest species and to secure the water catchment for the supply of drinking water to the town. All forest types are managed for maximum biodiversity retention.

Mangrove swamps are often filled in to reclaim land. However, they are immensely important ecosystems and play a vital role in the seafood industry, being nurseries for many kinds of larval fish and other seafood species.

• In the valley, the farmers maintain their remaining trees, fencing off small areas to encourage seedlings and saplings. Remnants of bushland are encouraged, especially along the river. Three things result: salt stops accumulating in the soil, populations of useful biological control agents begin to build up, and farmers see the return of wildflowers and butterflies. Several farmers form a Landcare group and contact a university research unit about an area by the river that has suffered particularly from overgrazing and soil erosion. The unit obtains a grant to establish a restoration project to reforest the area. There is general agreement that some timber can be taken by the farmer once the woodland is re-established.

• Down at the mangrove swamp, the developer hires a biologist who demonstrates how by building on only a small portion of the swamp the project can be advertised as an eco-tourist resort. It can offer selective fishing, boardwalks to see paperbark trees, orchids and insect-eating plants, boat tours to see the fantastic variety of colourful swamp birds and fish nurseries and, after dinners that include local bush-tucker, night-spotting to see the native marsupials. As a result, the developer becomes interested in maintaining local biodiversity.

• Meanwhile, in town, residents enjoy the parrots and recreational fishing. In consultation with the farmers and foresters, several reserves have been established to protect different kinds of vegetation. It has been agreed that over the years both the farmers and the foresters will allow some native vegetation to grow back to form corridors that link the reserves. Two enterprising families go bush to start a hotel that offers natural forest scenery and plenty of wildlife. A young couple start a biodiversity centre for local schoolchildren that also provides a seed bank and experimental plots for local farmers and foresters.

This kind of scenario may appear over-idyllic

Ecologically Sustainable Development

Wetlands are important centres of biodiversity and major tourist attractions. Crocodiles are keystone species that convey a sense of wildness which is an important feature for an increasing number of eco-tourists.

Ecologically Sustainable Development (ESD) has become a very important phrase in commerce, industry and politics but many people are uncertain what it means. In some ways it is very easy to understand because we often know what is not sustainable. For example, we know that most continuous intensive fishing is not sustainable because of the decimation of fish stocks. Many fisheries worldwide have been bankrupted by this process. We know that some kinds of forestry are not sustainable because they inhibit regrowth or, in some cases, because the loggers simply run out of trees. We know that some pastoralism is not sustainable because the land degrades under the influence of domestic stock, losing its vegetation, suffering catastrophic soil erosion. We know that some kinds of farming are not sustainable because the soil becomes excessively acid or salty. We know that some tourist developments are unsustainable because they destroy the very amenities the tourists come to see. We know that urbanisation may be unsustainable because the local ecosystems on which it depends become degraded and polluted with massive losses of biodiversity.

This much we know. With the benefit of hindsight we know that much of what we have done in the past was unsustainable. Most of the time nobody was to blame as people were just doing the best they could. However, now that we understand the dangers of the poor management of ecosystems when farming, logging, keeping stock, fishing, mining, constructing resorts or building cities, the only intelligent road ahead is to implement strategies that are sustainable.

To most people the word sustainable seems to mean economically sustainable, that is, providing long-term financial security. However, in all the examples discussed above, it is clear that economic sustainability relies heavily, and in some cases absolutely, on ecological sustainability. This is because the basic resources are genetic, species and ecosystem diversity and, if they are improperly managed, they will degrade, change or fail.

All this means that the Federal and State governments of the nation, together with our peak industries, are in the process of determining what changes should be introduced to make our industries and urban developments as sustainable—as secure—as possible. This requires long-term planning and general avoidance of the 'quick buck' motivation. A very practical aspect is bioregional planning. The basic aim is nothing less than to devise strategies that secure both the economic and ecological future of Australia.

but it demonstrates the immediate importance of biodiversity conservation at the community level. Preserving biodiversity as remnants, corridors and in reserves increases the sustainability of biodiversity-based industries by functioning as a basic resource for them. It also increases the quality of daily personal and community life by supplying recreation.

Bioregional management emerges from the needs and desires of local communities and the agreement of all of them is required, especially when there are conflicting points of view. Clearly, it is no longer acceptable to poison your neighbour's water, remove resources at rates faster than they can be replaced or destroy high amenity values in landscapes enjoyed by local communities and utilised by the tourist industry.

Bioregional management has the great advantage of being at about the right scale for action by most people; it makes sense to look outside local boundaries without going too far. Further, some of the necessary infrastructures may already exist, including local government associations, business groups, chambers of commerce and conservation societies. The role of big government may be mainly in the provision of information centres and expert advice when local problems appear insoluble. The universities and other research institutions where landscape-level ecology is being developed can already provide advice and know-how where appropriate.

The Murray–Darling Basin is an example of the integrated management required for the conservation of a large region. It incorporates one of the

How to Choose Areas for Reservations

Biodiversity is not evenly scattered across the surface of the Earth: some habitats support greater diversity than do others. Consequently a single habitat containing many species may be chosen as a biodiversity reserve. On the other hand, a bioregion may harbour many different habitats, which vary greatly in species diversity. Some may contain important genetic varieties and few species while others harbour rare species or communities. Therefore, the bioregion is often the appropriate unit for conservation because it harbours biodiversity at all three levels and because each habitat contributes something unique to the total biodiversity.

Suppose that a particular habitat is quite common but only a few can be reserved. How do you choose? Chris Margules and his team at the CSIRO have developed one approach to this problem, called the 'minimum set', the meaning of which will soon become clear. To test this method they worked with the wetlands that abound in the flood plain of the Macleay River and identified 432 separate wetlands containing 98 plant species, many confined to wetlands. They then asked the question: what is the minimum set of this kind of habitat that represents all species at least once, and hopefully more times than this?

Their search showed that 20 wetlands were required to represent all species at least once. Those 20 wetlands represented 45 per cent of the total area under consideration but they varied greatly in size. They also showed that creating greater security for these wetland species by having each represented five times, required 65 wetlands or 78 per cent of the area available. They concluded somewhat pessimistically that, 'The conservation of even one population of every plant species in these wetlands is therefore unlikely' *(Margules and Austin, 1991).*

most extensive production areas in Australia, as well as significant conservation reserves and water resources but it suffers from severe soil and water degradation. These resources were originally managed by many independent statutory authorities from several States. Decisions made by one authority often had far-reaching, adverse consequences on resources elsewhere in the region. A single authority was formed in order to ensure a coherent management policy governing both private and public land. This resulted in better communication between the agricultural, vegetation and soil managers.

Protected Areas

To assist in the process of reconciling the protection of biodiversity with the use of land for production, several categories of protected areas have been suggested by the International Union for the Conservation of Nature (IUCN), now known as the World Conservation Union.

Fully protected areas: these are managed to strictly limit human settlement, cultivation and invasion by exotic species and include several types:

• Strict nature reserves are generally small areas set aside to conserve some important natural phenomenon such as a cave, forest remnant, the headwaters of a river or an estuary.

• National parks are usually much larger and contain a variety of landscapes and ecosystems that people may visit for education, recreation, inspiration and scientific research provided that biodiversity is not threatened.

• National monuments are designated in several countries to protect a single spectacular natural feature or historic site.

Extractive protected areas include:

• Habitat and wildlife management areas, which are managed not only to encourage and protect wildlife but to use it where appropriate. A special category developed overseas is the extractive reserve, which is modelled on the original methods of harvesting natural resources by indigenous people. Tropical forests yield hundreds of products including fruits, nuts, fibres, resins, oils and pharmaceuticals that may exceed the value of the timber. So important is this concept that 10 per cent of Brazilian forest under government control is dedicated as extractive reserves. It does not mean that no timber is taken but it becomes just one of a wide variety of products. When properly managed, such products can provide a large and sustainable income without ecologically significant damage to the forest.

• Protected landscapes are likely to be a mix of private and public land where biodiversity is subject to resource extraction such as farming, forestry and fishing, and to harbour human settlements of all kinds. Here the objective is to practice bioregional planning to integrate, as far as possible, the objectives of commercial production, water catchment management and the protection of biodiversity.

Other types of reserves include State recreation areas, State forests, flora reserves, wildlife refuges, travelling stock routes, Land for Wildlife (Victoria), private recreation reserves, municipal parks and marine parks. It is important to note that none of these were established to conserve biodiversity as we understand the term today.

Over the last quarter century, protected areas have been established at a much faster rate. On a worldwide scale, between 1900 and 1970, reserves larger than 1,000 hectares were set up at a rate of approximately 22,000 square kilometres per year. By 1970, these protected areas totalled roughly 1.6 million square kilometres, almost the size of Queensland. In the 20 years since, this rate has increased almost ten-fold to a total of seven-and-a-half million square kilometres. This would nearly cover Australia but represents only one twentieth of the land surface of the Earth.

Currently a total of 5.3 per cent of the land

area of Australia is reserves, which is equivalent to about half the size of New South Wales. Thirty-five per cent of these are under strict protection whereas some form of resource extraction is permitted in the remaining 65 per cent. On a worldwide scale, 58 per cent of such areas are under strict protection.

In Australia, only one fifth of endangered and vulnerable plant species occur in national parks or proclaimed reserves. Perhaps with time, this proportion will be significantly increased but until then the importance of non-reserved areas for the conservation of biodiversity is clear. We know that many species, both plant and animal, persist in such places as remnants of natural vegetation on farms, patches of unlogged forest on steep hillsides, suburban areas with gardens and abandoned industrial yards. Increasing numbers of Australians are very active in maintaining the biodiversity of these non-reserved areas. However, they can be greatly increased in importance by the management of the areas that surround them in ways that are sympathetic to biodiversity. Many new practices in logging, grazing, fish harvesting and urbanisation have been developed for this purpose.

In some parts of the world regional consultations have resulted in the creation of large areas where local economies have been adjusted to meet the dual objectives of sustainability and biodiversity conservation. These economies are mostly rural and some are indigenous. Parts of East Africa where cattle are grazed support as many game species as other areas within national reserves. The main factor affecting the diversity of game species appears to be rainfall rather than the presence or absence of cattle. However, social factors are crucial to prevent this situation from breaking down. Of greatest importance is the maintenance of a stable culture and security of land tenure so that over grazing does not occur. In addition, game animals are valued for their meat and other products and are managed, along with the cattle, for sustainable use.

Coastline of Yuraygir National Park in northern New South Wales. Roughly one-twentieth of Australia is reserves and of these a third are under strict protection. However, much biodiversity lies outside reserves.

Some species are failing to persist anywhere. This may be the result of a combination of factors such as drought and poaching. It has often been possible to maintain small populations of rare or endangered species in zoos or botanic gardens. However, the maintenance of plants and animals in captivity can be expensive and the number of species involved is consequently very limited. Reliance on captive breeding to save a species is almost always the last resort when they appear to be doomed in the wild.

Conservation Reserve Design

In Australia, as in most of the world, large areas of habitat have been cleared leaving 'islands' of remnant habitat surrounded by 'oceans' of agriculture, urbanisation, logged forests, and other kinds of development. In many cases these 'islands' are all that is left to conserve and there is no choice about the design of reserves. However, land managers are increasingly aware of the need to plan for reserves and a number of major design issues have emerged.

The Size and Number of Reserves

There is widespread support for the principle that larger reserves will retain more genetic variation, species and ecosystems than smaller reserves. For example, research on mammals in western Victoria showed that, as reserve size increased, the number of species each contained increased. The picture for birds is more complex. Studies in remnant vegetation in Victoria have indicated that reserve size is not the only factor. When the understoreys of remnants were destroyed by grazing cattle, Noisy Miners moved in and actively attacked other birds, reducing bird diversity. Revegetating the shrub layer reversed this trend. When the Noisy Miners retreated, the other native birds returned.

Rocks covered with mosses in wet sclerophyll forest, New South Wales. These humble plants are rarely considered in conservation strategies.

While size can be important it has been argued that establishing a larger number of smaller reserves can achieve similar goals. This dilemma has been given the acronym SLOSS: Single Large Or Several Small? Numerous small reserves can buffer species against chance extinction better than one large one. A disease could eliminate a species in a single reserve but the species may survive the attacks of disease when it has genetically distinct populations scattered in several places. This has been significant in Koala conservation.

It seems likely that the importance of reserve size varies according to the habitat and species being considered. For example, studies of lizards in the wheat belt of Western Australia have shown that a number of small reserves provides a greater variety of habitats than it is possible to find in a single reserve unless it is enormous. Enormous reservations are simply unavailable. However, other studies confirm that groups of small reserves are not capable of supporting large or wide-ranging species. Herds of migratory herbivores, animals with big territories and species such as predatory birds that hunt across huge areas therefore require major reservations for survival.

Is Reserve Shape Important?

Reserves with small boundary-to-area ratios retain more species than those with long or irregular boundaries. Many more forest bird species will live in large circular or square blocks of forest than in long thin ones, such as roadside avenues. In addition, when the boundary-to-area ratio is high, the effects of the reserve surroundings can be magnified by the so-called 'edge effect'. For example, a long perimeter may be subject to more urban or industrial run off and may increase the area subject to invasion by exotic weeds.

Irregularly shaped reserves can be made more useful by the creation of surrounding buffer zones in which the land is managed in a manner sympathetic to biodiversity. Some farmers, for example, tolerate kangaroos and other wildlife in the paddocks that surround nature reserves and national parks.

How Useful are Wildlife Corridors?

It is often argued that by linking reserves with corridors of native habitat, their effective size is greatly enhanced thus protecting more species. However, so few experiments have been performed to check on this we do not know whether or not corridors are generally useful. Woodland mice and chipmunks in Canada and some bird species in the wheat belt of Western Australia have been shown to use them. Hedgerows and roadside reserves may function as corridors and these are also thought to be important to some kinds of small mammals and birds. However, the effectiveness of corridors for smaller or less mobile organisms such as lizards, frogs, insects and plants is not well understood.

Unless carefully managed, corridors could instead become a problem, acting as conduits for weeds, predators, pests and diseases. Also, some species do not always behave in the same way in different places. For example, Koalas in suburban Ballarat ignored wide corridors while in the Gold Coast region of south-eastern Queensland they use corridors as narrow as avenues of roadside trees.

It is likely that many organisms simply do not perceive corridors in the same way that we do. For example, some butterflies and birds may respond more to the open space on either side of a corridor than to the corridor itself and simply turn away from it. We have no information on the way native plants may be able to utilise corridors. It would seem to be a slow process. For reasons such as these, corridors may be useful for some organisms and not for others and so they are designed with particular species or groups of organisms in mind.

Importance of Keystone Species

A keystone species is often defined as one that can help maintain diversity by sustaining other species during stressful times. For example, South American figs feed many species of birds and monkeys during drought. A keystone species also creates habitats for others and maintains their populations in such a way that its loss would lead to a loss in diversity. Sometimes these species are large or charismatic and reserves are often designed to maintain them. One example is the elephant in East Africa, which helps maintain the savanna habitat by pruning or removing trees. In some areas of Australia, spinifex grasses are keystone species. In Antarctic waters, krill are an essential link in the food chain of many animals, such as whales.

Disturbance and Reserves

Natural disturbance is a major cause of change in all ecosystems. In forests, for example, trees fall and create spaces where smaller and more short-lived species thrive. The banks of rivers cave in after floods, a disease wipes out a population of shrubs, a pollinator fails to appear for a few seasons and the plants it services go into a decline or a fire blackens a hillside. These kinds of processes contribute to overall diversity by allowing many different species to gain a foothold. Thus, a forest is a mosaic of different patches of vegetation and their dynamics are important to biodiversity.

Ecosystems change through time so that even wilderness is slowly evolving. Evidence from the past, especially fossil pollen grains that have fallen in layers over millions of years, show that on the scale of centuries, what must have seemed to be very permanent forests or grasslands were slowly transformed into different ecosystems. The problem today is that the disturbances caused by humans are too frequent, widespread and rapid. The natural restorative processes, and evolution itself, cannot keep pace.

Reserves and Climate Change

Reserves will be most effective if they are designed with the needs of migrating species in mind. As the greenhouse effect is expected to be greatest at mid-to-high latitudes, reserves are going to be especially important in the temperate zones and at higher lat

The Option Value of Biodiversity

At present the value of most species is unknown, however, research is revealing these values at an ever-increasing rate. Even the most unexpected groups of organisms, such as bacteria, worms, ants and spiders, have immense value, either as parts of their ecosystems or as sources of commercial products—or both (see Chapter 3).

Some economists now take account of our current ignorance of the value of species by giving them what has become known as option value. This means that although this generation cannot assign a value to a species, future generations are likely to have developed the technologies, life-styles and ethics that make the value of each species very clear. Option value may be entirely economic, anticipating that a species will gain value as a commodity. It may also be aesthetic or it may be because a species is an important part of an ecosystem.

The key to the option value system is that the species must still exist. Clearly if a species is driven to extinction, the option is ended and the value lost. The value of such species could be direct—for the provision of vital needs such as food, medicine, clean water—or indirect, like attracting tourists. If we were to deny such resources to a group of people in a different place, this would be considered an act of war. The denial of such resources to people in a different time—our own children and their descendants—is equally unpardonable.

itudes. Large reserves, or systems of reserves with migration corridors along altitudinal and latitudinal gradients may be the most effective. Some extremely thorny questions may arise as the climate warms, for example, should species that cannot migrate with sufficient speed be transplanted by conservationists? If so, which ones?

Ecosystem Restoration

Disturbance of land is an inevitable consequence of civilisation and the world is covered with ecosystems that have been degraded in many different ways. This degradation has been recorded by historians for centuries so it is surprising that a scientific discipline to understand and combat it has only recently emerged. However, restoration ecology is now one of the fastest growing of the biological sciences as countries all over the world scramble to try to correct the ravages of centuries of exploitation.

Ecosystem restoration is basically a problem of biology, although there may be major engineering phases. For example, garbage dumps and mine tailings may initially require vast amounts of soil to be imported by truck and its subsequent levelling by bulldozers to form a bed suitable for vegetation. This is where the story really begins as the restoration of native vegetation and its normal fauna requires detailed understanding of the biology of the species involved.

Nowhere is this more obvious than in the gigantic city garbage dumps of New York. Located on nearby Staten Island they are so huge that they can be seen on satellite photographs. Once soil was brought in, Steven Handel, the restoration ecologist in charge, had the daunting task of coaxing local species to grow. If he chose grasses, there would be little shelter for other species to take root. If he chose trees there might be little coverage for years. Instead he chose local shrub species that produced berries. The trick worked as within three years the shrubs were established and attracting native birds, which sheltered

Carnivorous plants, such as these Western Australian species, digest insects to augment an inadequate supply of nutrients. Carnivory is an adaptation for survival in nutrient-poor soils.

in their branches and ate their fruits. During their visits seeds already in their guts came out in their droppings. The following season the seeds germinated to form the beginnings of a new community. While only 10 species were planted originally, many more were brought in by the birds. New Yorkers are now acutely aware that the greatest resource the city has to restore its garbage dumps is bird biodiversity!

Australia is leading the world in many kinds of restoration technology, especially for mine sites. Jonathan Majer of Curtin University in Perth and the environmental officers of many mining companies are beginning to understand how to re-assemble species into communities that have a chance to grow, develop and rebuild local biodiversity. This is a major advance over previous techniques, which usually involved merely covering the site with a layer of green, often exotic grasses or pine trees. These non-native plantations often failed to persist, sometimes because of fire (see box on page 90), or lingered on with minimal biodiversity. Of course, old mine sites are small relative to many areas of degraded forest and farmland. The restoration of much larger areas remains a challenge to the creativity of restoration ecologists.

Biodiversity is the resource for restoration. Today, ecologists recognise this and select local plant species that can stand the harsh conditions of a degraded site and persist long enough to gather more species around them. In turn, they are the most likely to attract the local insects, birds and mammals to form a new community and to establish the complex biodiversity that new soils require to remain fertile. Biodiversity is also used as an indicator of the status of restoration. When species or groups of species that were wiped out by human activity return and successfully reproduce, then the project may be on the road to success.

This is an appropriate note to end this book because it shows that although biodiversity has been severely battered by centuries of human activity, it is also the very resource that can help us repair the damage. Biodiversity, when properly managed, can help restore the quality of our air, water and soils. It can reduce reliance on chemicals in agriculture and sustain pastoralism, forestry, fisheries and tourism. And, in all those places where ecosystems have been degraded or destroyed, it provides the raw materials for restoration and rehabilitation. On top of all this, biodiversity provides us with beauty and recreation. Life has not yet been discovered elsewhere in the universe but we are not alone as the biodiversity of Earth provides us with astonishingly productive and inspirational company.

Suggested Reading

Adamson, D.A. and Fox, M.D., 1981.
Change in Australian vegetation since European settlement. In: **A history of Australian vegetation,** J.M.B. Smith (ed.), pp. 107—147.
McGraw-Hill: Sydney.

Barr, N. and Cary, J., 1992.
Greening a brown land: the Australian search for sustainable land use.
Macmillan Education: Melbourne.

Commonwealth Government, 1991.
Ecologically sustainable development working groups final report and executive summaries.
Australian Government Publishing Service: Canberra.

Margules, C.R. and Austin, M.P., 1991.
Nature conservation: cost effective biological surveys and data analysis. CSIRO: Canberra.

Moritz, C., Kikkawa, J. and Doley, D., 1993.
Conservation biology in Australia and Oceania.
Surrey Beatty & Sons: Chipping Norton.

Morton, S.R., 1990.
The impact of European settlement on the vertebrate animals of arid Australia. A conceptual model.
Proceedings of the Ecological Society of Australia. 16: 201–213.

Peters, R.L. and Lovejoy, T.E., 1992.
Global warming and biological diversity.
Yale University Press: London.

Saunders, D.A., Arnold, G.W., Burbridge, A.A. and Hopkins, A.J.M., 1987.
Nature conservation: the role of remnants of native vegetation.
Surrey Beatty & Sons: Chipping Norton.

Saunders, D.A.,Hopkins, A.J.M. and How, R.A.,1990
Australian ecosystems: two hundred years of utilization, degradation and reconstruction. In **Proceedings of the Ecological Society of Australia,Vol.16**
Surrey Beatty & Sons: Chipping Norton.

Solbrig, O.T., Van Emden, H.M. and Van Oordt, P.G.W.J., 1992.
Biodiversity and global change.
International Union of Biological Sciences: Paris.

Western, D. and Pearl, M., 1989.
Conservation for the twenty-first century.
Oxford University Press: Oxford.

Glossary

adaptation A particular structure, physiological property or behaviour that enhances survival and reproduction in a particular environment.

aerobe Any organism that can survive only in the presence of oxygen.

anaerobic Ability to live in the absence of (gaseous or dissolved) oxygen.

arthropods A phylum of animals that possess jointed legs and a hardened outer skeleton; including insects, crustaceans and spiders.

autotroph An organism that derives its energy directly from the physical environment, such as sunlight or inorganic chemical reactions. See heterotroph.

base A chemical substance that has a tendency to accept protons (H+); a base dissolves in water with the production of hydroxyl ions and reacts with acids to form salts.

biodiversity The variety of life forms: the different plants, animals and micro-organisms, the genes they contain and the ecosystems they form. It is usually considered at three levels: genetic diversity, species diversity and ecosystem diversity.

biological control Control of pests by the use of organisms or their products. Two common methods: 1. to increase the numbers of the predators, parasites or pathogens of the pest; 2. to bait traps with naturally attractive chemicals such as the sex attractants of insect pests.

biome A major regional community of organisms usually defined by the vegetation and determined by the interaction of the substrate, climate, fauna and flora. The term is usually limited to terrestrial habitats, e.g. eucalypt forest, desert. Oceans and bodies of fresh water may also be considered to be biomes.

bioregion A territory defined by a combination of biological, social and geographic criteria rather than by geopolitical considerations.

biota All the living creatures: plants, animals and micro-organisms, in an area.

blue–green algae Photosynthesising bacteria of the group known as Cyanobacteria.

bromeliads A large family of herbaceous plants, mostly from the tropical Americas, including pineapples and a large number of ornamental species. Many are found as epiphytes on larger plants and trees.

cell The structural unit of most organisms bound by a membrane and containing genetic material, and the molecules and structures necessary for carrying out life processes.

chlorophyll Green pigments having the vital function of absorbing light energy for photosynthesis.

chromosome The structure by which hereditary information is physically transmitted from one generation to the next; the organelle that carries the genes. In bacteria, the chromosomes consist of a single naked circle of DNA; in eukaryotes, they consist of DNA molecules and associated proteins.

consumer Any organism that lives by feeding on other organisms, dead or alive. The term includes all animals–herbivores, carnivores, and decomposers; parasitic and decomposer plants; and most micro-organisms. Those left are producers.

continental drift The continuous motion of the continents on moving portions of the Earth's crust known as tectonic plates.

convergent evolution The independent development of similar structures in organisms that are not related; generally found in organisms living in similar environments.

corridors Also known as habitat corridors: strips of land of varying width between reserves to facilitate migration, colonisation and interbreeding.

CSIRO Commonwealth Scientific and Industrial Research Organisation.

culture Micro-organisms grown on or in a nutritious medium, usually in a glass dish or test tube.

decomposer An organism that survives by feeding on the carrion or wastes of other organisms. Decomposers play a crucial role in nutrient cycles by breaking down complex organic molecules into simpler constituents.

dinoflagellate Unicellular photosynthetic organisms forming part of phytoplankton.

DNA (deoxyribonucleic acid) The genetic material of all organisms; composed of two complementary chains of nucleotides wound in a double helix.

dormancy A special condition of arrested growth in which a plant and such plant parts as buds and seeds do not begin to grow without special environmental cues.

ecological sustainability Meeting the needs of present and future generations with minimal damage to the environment or loss of biodiversity.

ecosystem The organisms in a community plus the associated abiotic (non-living) factors with which they interact.

ecosystem services Services provided to humanity by natural ecosystems such as amelioration of weather, control of water flows, maintenance of soils, disposal of wastes, and recycling of nutrients.

endemic Restricted to a specified region or locality.

epiphyte Plants that grow attached to other plants solely for support.

eukaryote A cell characterised by membrane-bound organelles, most notably the nucleus, and by chromosomes whose DNA is associated with proteins.

evenness (equitability) The distribution of numbers of individuals of each species in a community. For example, a community of 100 individuals may have 20 belonging to each of five species (very even) or 80 belonging to one species and five belonging to each of four other species (very uneven).

evolution The continuing process of change and diversification that has led, over billions of generations, from one or a few simple proto-organisms to the rich array of life forms that have populated Earth.

exoskeleton An external skeleton, as in arthropods.

extinction The loss of all members of a species.

family A taxonomic category between genus and order. Dogs and their relatives (wolves, jackals, foxes) are in the family Canidae; cats and their relatives (lynxes, lions, tigers, cheetahs) are in the family Felidae.

fermentation A process by which some micro-organisms gain energy under conditions where oxygen is not present.

flagellum A fine, hair-like extension of a cell, associated with locomotion in single-celled organisms.

food web The food relationships between species within a community. A diagram of who eats whom.

Foraminifera Amoeboid protozoans that possess chitinous, hard shells that are usually many-chambered.

gene The fundamental physical unit of heredity coded in DNA molecules; it transmits information from one cell to another and hence one generation to another.

genus (pl. genera) A set of one or more species; the main taxonomic category between species and family. Humans belong to the genus *Homo*.

greenhouse effect Atmospheric warming when solar energy is absorbed by gases such as carbon dioxide. The heat is then trapped within the atmosphere. An enhanced greenhouse effect, and therefore an increase in average global temperature, is probably occurring because of increased carbon dioxide from burning of fossil fuels.

habitat The environment of an organism; the place where it lives.

heathland Open uncultivated land covered by low, usually small-leaved shrubs.

herbaceous A general term referring to any non-woody plant.

herbarium A scientific collection of dried and pressed plant specimens.

herbivore An organism that obtains nutrition and energy by eating plants, e.g. grasshopper, kangaroo.

heterotroph An organism that cannot derive energy from photosynthesis or inorganic chemicals and so must feed on other organisms obtaining chemical energy by degrading their organic molecules. Animals, fungi, and many unicellular organisms are heterotrophs; see also autotroph.

host-specificity The degree to which a herbivore or parasite has become specialised in exploiting host species. Highly specific herbivores and parasites use only one host species at any particular stage of their life-cycle.

humus Decomposing organic matter in the soil.

hypha (pl. hyphae) A filament of a fungus. Collectively the hyphae comprise the mycelium.

invertebrate Any animal that does not possess a backbone, for example, insects, worms, molluscs.

kingdom The highest grouping (taxon) in biological classification.

lithotrophs Organisms that gain energy from inorganic sources, e.g. hydrogen sulphide (rotten egg gas) or ammonia.

megadiverse An informal concept that describes a small number of countries, mostly tropical, that contain very high levels of biodiversity and endemism. They include Brazil, Colombia, Ecuador, Peru, Mexico, Zaire, Madagascar, China, India, Indonesia, Malaysia and Australia.

methane A simple organic gas (CH_4) found in natural gas and produced by some bacteria.

microclimate The climate in the immediate vicinity of an organism or in a specific habitat.

micro-organisms Microscopically small organisms, including single-celled plants and animals, bacteria and many fungi. Viruses are commonly included.

microbes Micro-organisms.

molecular biology The study of the structures and properties of complex molecules found in cells, particularly DNA, RNA and proteins.

molecule A collection of two or more atoms held together by chemical bonds; the smallest unit of a compound that displays the properties of the compound.

mutation A change in the genetic material of an organism.

mycelium The total mass of hyphae of a fungus.

natural selection The mechanism by which gradual evolutionary changes take place. Individuals that are better adapted to the environment in which they live produce more viable young, thus increasing their proportion in the population.

nematodes Roundworms. Abundant in most environments, including many forms parasitic on animals, plants and fungi.

nitrate-reducing bacteria Bacteria that use oxidised forms of nitrogen instead of oxygen for respiration.

notochord A flexible dorsal rod that runs the length of the body in chordates and forms the primitive axial skeleton in the embryo. In most adult chordates the notochord is replaced by a vertebral column that forms around the notochord.

nucleus An organelle of eukaryotic cells that is enclosed by a nuclear membrane and contains the chromosomes.

nutrient A substance necessary for the normal growth, development and reproduction of an organism. Often used in the more restricted sense of inorganic nutrients taken up by plants from air or water.

order A major category above the family and below the phylum. Butterflies and moths belong to the order Lepidoptera, wolves and lions to the order Carnivora.

organelle A well-defined structure within a cell, e.g. nucleus.

organic Refers to living organisms in general, to compounds formed by living organisms or to the chemistry of compounds containing carbon.

ovule A structure containing an egg in seed plants.

parasite An organism that lives in association with, and at the expense of, another organism, the host, from which it obtains organic nutrition.

parasitoid Free-living insects whose larvae grow and develop in a host, not killing it until they emerge as adults.

passerine An order of birds, including more than half of all known species. Typically these are perching birds including finches, warblers, swallows, crows.

pathogen Any organism that causes disease, usually a micro-organism.

PCBs (polychlorinated biphenyls) Industrial chemicals so stable that they accumulate in the environment and pass through food chains. Natural micro-organisms are being used to find ways of breaking them into harmless substances such as carbon dioxide and water.

photosynthesis The conversion of light energy to chemical energy in living organisms.

phytoplankton That part of the plankton made up of plant life.

placental A common term applied to mammals that nourish their developing embryos via a placenta and give birth to well-developed young (as opposed to the pouched marsupials and the egg-laying monotremes).

plankton Free-floating or swimming, mostly microscopic, freshwater and marine organisms.

producer A green plant or other organism that can bind energy from inorganic sources into the chemical bonds of organic compounds. See consumer.

prokaryote A bacterium; a cell lacking a membrane-bound nucleus or membrane-bound organelles. Prokaryotic cells are more primitive than eukaryotic cells, which evolved from them.

protein A complex organic compound made up of many amino acids linked together.

protozoa Single-celled animal-like organisms that live either singly or as colonies.

rangeland Open uncultivated land usually covered by grasses, herbs and shrubs, used to pasture domestic animals which often coexist with native grazers.

refugium (pl. refugia) Locality that has escaped drastic alteration following climatic change such as glaciation, in contrast to the region as a whole. Refugia usually form centres for relict species, populations or communities.

relict Surviving population or community characteristic of an earlier time. A relict distribution of fauna or flora is one representing the localised remains of an originally much wider-ranging distribution.

RNA (ribonucleic acid) A class of nucleic acids characterised by the presence of the sugar ribose (DNA contains deoxyribose) and the pyrimidine uracil (DNA contains thymine).

sclerophyll (pl. refugia) Types of plants, often shrubs, with specialised, tough leaves to reduce water loss.

speciation The process by which new species are formed.

species (pl. species) A group of organisms that actually (or potentially) interbreed in nature; a taxonomic grouping: the category below genus.

springtail An order (Collembola) of small to minute wingless insects found in moist places such as leaf litter, soil and rotten wood.

stamen The part of the flower producing the pollen, composed (usually) of anther and filament.

stigma The receptive surface for pollen grains.

streptomyces A group of filamentous soil bacteria that are responsible for the characteristic smell of soil. These bacteria are the most important producers of antibiotics.

stromatolites Rock-like structures containing alternate layers of blue–green algae (Cyanobacteria) and marine sediments. They may be either living or fossil, the latter formed up to 3,000 million years before present.

sulphate-reducing bacteria Bacteria that use oxidised sulphur compounds instead of oxygen (in the air) for respiration. These bacteria generate hydrogen sulphide, which is frequently smelt around marshy areas.

sulphur bacteria A type of lithotroph that gains energy from inorganic sulphur sources.

taxonomy The science of the classification of organisms.

thermophiles Organisms that live in hot environments and grow best at temperatures above 45°C.

tundra Treeless region extending to the farthest limit of plant growth. A huge biome, occupying about one fifth of the land surface, found mostly north of the Arctic circle. Usually permanent ice lies within a metre of the surface. Bryophytes and lichens are common organisms.

wetlands An area in which the soil is permanently or regularly saturated with water, e.g. swamps and marshes.

Index

This index was prepared by Jonathan Jermey, a registered member of the Australian Society of Indexers.

References to illustrations are shown as italic page numbers: e.g. cicada *40.*

Photographic Credits

Preface
7 "Stalk-eyed fly" – Sue Linsey/Australian Museum

Chapter one
10 "Great Barrier Reef" – Kathie Atkinson
11 "Cushion Plants" – Kathie Atkinson
12 "Longicorn Beetle" – Pavel German
13 "Foraminifera" – Ron Oldfield
"Slime Moulds" – Ron Oldfield
14 "Cassowary" – Hans & Judy Beste/Auscape International
17 "Gouldian Finch" – Andrew Henley/Auscape International
18 "House dust" – The Photo Library-Sydney/David Scharf/SPL
20 "Ecosystem Variety" – Kathie Atkinson

Chapter two
22 "Diatoms" – The Photo Library-Sydney/Manfred Kage/SPL
23 "Dickinsonia Fossil" – Jim Frazier
25 "Stromatolites" – Reg Morrison/ Auscape International
27 "Clingfish" – Barry Hutchins
29 "Albatross" – Kathie Atkinson
32 "Cradle Mountain" – Kathie Atkinson

Chapter three
34 "Green Ants" – Kathie Atkinson
36 "Native Plants" – Vic Cherikoff/Bush Tucker Supply Australia
37 "Heron Island" – Kathie Atkinson
41 "Cotton Farmer" – Auscape International
43 "Toxic Algae" – Kathie Atkinson
44 "Spider Silk" – Kathie Atkinson
45 "Mussels" – Kathie Atkinson

Chapter four
46 "Sea Slugs" – Bill Rudman/Australian Museum
47 "Slugs" – Kathie Atkinson
48 "Beach and Ribbon Worms" – Kathie Atkinson
49 "Octopus" – Kathie Atkinson
50 "Peripatus" – Kathie Atkinson
51 "Tardigrada" – Ron Oldfield
52 "Cycads" – Kathie Atkinson
53 "Fern" – Kathie Atkinson
54 "Flying Fox" – Kathie Atkinson
55 "Waterlily" – Rod Peakall

Chapter five
56 "Proteus mirabilis" – The Photo Library/A B Dowsett/SPL
58 "Hydro Vents" – The Photo Library/PeterRyan/SPL
59 "Volvox" – The Photo Library/Manfred Kage/SPL
61 "Pea Plant" – The Photo Library-Sydney/Dr Jeremy Burgess/SPL
62 "Fungal Hyphae" – The Photo Library-Sydney/David Scharf/SPL
63 "Lichen" – Jiri Lochman/Lochman Transparencies
65 "Protozoa" – Ron Oldfield
66 "Saline Lake" – Jiri Lochman/Lochman Transparencies

Chapter six
68 "Penguins" – Graham Robertson/Auscape International
71 "Coolibah Trees" – Kathie Atkinson
72 "Butterfly" – J. Cancalossi/Auscape International
73 "Lord Howe Island" – John & Val Butler/Lochman Transparencies
74 "Nudibranch" – Kathie Atkinson
77 "Deep Sea" – Alistair Birtles
78 "Seaweed" – Kathie Atkinson

Chapter seven
80 "Nothofagus" – Dennis Harding/Auscape International
82 "Protea" – Rod Peakall
84 "Wattle" – Kathie Atkinson
86 "Frogs" – Kathie Atkinson
87 "Thorny Devils" – Kathie Atkinson
88 "Eucalypt" – C. Henley/Auscape International
91 "Orchid" – Rod Peakall
92 "Mole" – Mike W. Gillam/Auscape International
93 "Spinifex" – Jean-Paul Ferrero/Auscape International

Chapter eight
94 "Fiddler Beetle" – Brett Gregory/Auscape International
95 "Biologist" – Robyn Stutchbury
97 "Salvinia molesta" – CSIRO
100 "Escherichia coli" – The Photo Library-Sydney/Photo Researchers Inc.
101 "Mites" – Mike Gray/Australian Museum
102 "Coral Reef" – Eva Boogaard/Lochman Transparencies

Chapter nine
106 "Corroboree Frog" – Jean-Paul Ferrero/Auscape International
108 "Bilby" – Kathie Atkinson
111 "Land Degradation" – Kathie Atkinson
113 "Byron Bay" – Jean-Paul Ferrero/Auscape International
114 "Mangroves" – Kathie Atkinson
115 "Wetlands" – Auscape International
118 "Yuragir National Park" – Kathie Atkinson
120 "Mosses" – Kathie Atkinson
123 "Pitcher Plant" – Marie Lochman/Lochman Transparencies